EXAMINATION TEST PAPERS FP1/FP2/FP3

The test papers are based on the EDEXCEL Syllabus but they would be very useful for OCR, AQA and other examining boards.

CONTENTS

There are 10 (1 hour and thirty minutes) FP1 Test Papers, 10 (1 hour and thirty minutes) FP2 Test Papers and 10 (1 hour and thirty minutes) FP3 Test Papers with marking scheme and the full solutions.

(Anthony Nicolaides)

ISBN-13 978-1-872684-50-5 £19.95 Book

ISBN-13 978-1-872684-52-9 £19.95 CD

Examination Test Papers.

GCE Advanced Level.

Further Pure Mathematics FP1

Test Paper 1

Time: 1 hour and 30 minutes.

Instructions and Information

Candidates may use any calculator allowed by the regulations of their Examination Board.

Full marks are awarded for correct answers to ALL questions.

This paper has nine questions.

You can start working with any question and you must label clearly all parts.

1. Sketch the parabolas:

(i) $y^2 = 4ax$ (ii) $y^2 = -4ax$ (iii) $x^2 = 4ay$ (iv) $x^2 = -4ay$

which are curves produced by intersecting cones by planes parallel to the slants.

Consider the values $\left\{ \begin{array}{llll} x = a, 4a & \text{and} & x = -a, -4a & x = 0 \\ y = a, 4a & \text{and} & y = -a, -4a & y = 0. \end{array} \right\}$ **(8)**

---

2. (a) Sketch the graph of $y = x - 2$ and the graph $y = \dfrac{1}{|x+2|}$ on the same diagram. **(3)**

Indicate on your sketch the coordinates of any points at which the graph crosses the axes and state the equations of any asymptotes. **(4)**

(b) Find the set of values of x for which $x - 2 < \dfrac{1}{|x+2|}$. **(4)**

---

3. If $\mathbf{A} = \begin{pmatrix} 3 & -1 \\ -4 & 3 \end{pmatrix}$ and $\mathbf{B} = \begin{pmatrix} \frac{3}{5} & \frac{1}{5} \\ \frac{4}{5} & \frac{3}{5} \end{pmatrix}$.

Find (i) **AB** (ii) **BA**.

Comment on the results. **(6)**

---

4. Find (i) the moduli and (ii) the arguments of the complex roots of the quadratic equation $z^2 - 4z + 8 = 0$. **(4)**

---

5. The quadratic function is shown in Fig. 1

Determine the equation of this function. **(6)**

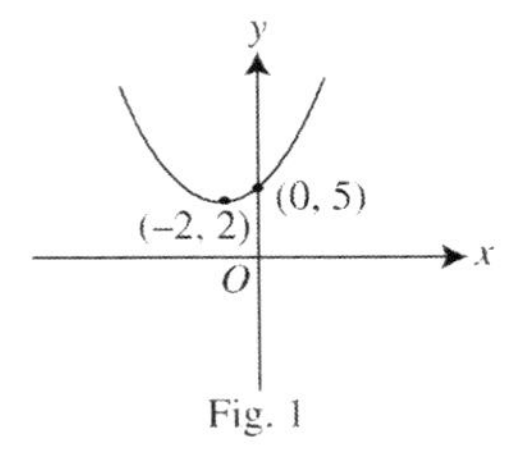

Fig. 1

---

6. The vertices of a parallelogram are given by A' $(-3, -4)$, B' $(3, -2)$, C' $(2, -6)$ and D' $(-4, -8)$.

Determine the coordinates of the reflected parallelogram in the x-axis. **(4)**

---

7. (a) If α and β are the roots of the quadratic equation $ax^2 + bx + c = 0$, show that the quadratic equation can be written in the form of $x^2 - Sx + P = 0$ where $S = \alpha + \beta$.

Hence show that $S = -\frac{b}{a}$ and $P = \frac{c}{a}$. **(5)**

(b) Use the method of completing the square that the roots are

$\alpha = \frac{-b + \sqrt{b^2 - 4ac}}{2a}$ $\quad \beta = \frac{-b - \sqrt{b^2 - 4ac}}{2a}$ and hence show that

$S = -\frac{b}{a}$ and $P = \frac{c}{a}$. **(7)**

---

8. Show that $\sum_{r=1}^{n} r = 1 + 2 + 3 + ... + r + ... + n = \frac{n(n+1)}{2}$ by using

(i) an arithmetic series **(5)**

(ii) the difference of squares. **(7)**

---

9. (a) Find the roots of the quadratic equation $z^2 + z + 1 = 0$, **(2)**

giving your answer in the form $a + ib$. **(2)**

(b) Show these roots z_1 and z_2 on an Argand diagram.

(c) Find (i) the sum of the roots, **(1)**

(ii) the product of the roots **(1)**

(iii) $\frac{z_1}{z_2}$

(iv) $\frac{z_2}{z_1}$ showing that the roots in (iii) and (iv) are conjugates. **(6)**

TOTAL FOR PAPER: 75 MARKS

Examination Test Papers.

GCE Advanced Level.

Test Paper 1 Solutions

Further Pure Mathematics FP1

1. (i)

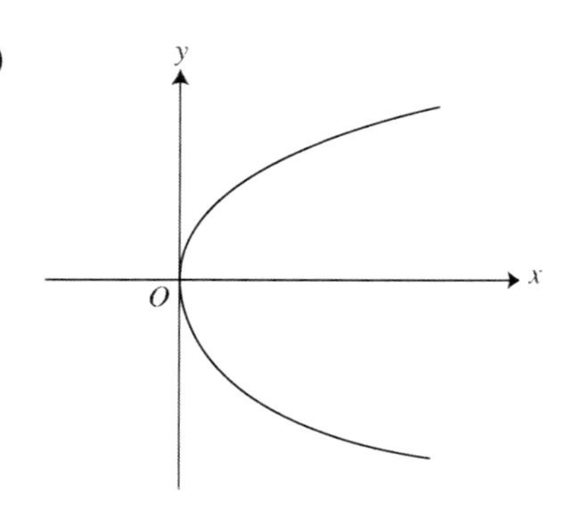

$y^2 = 4ax$

if $x = a$, $y^2 = 4a^2 \Rightarrow y = \pm 2a$

if $x = 4a$, $y^2 = 16a^2 \Rightarrow y = \pm 4a$

if $x = 0$, $y = 0$.

(ii)

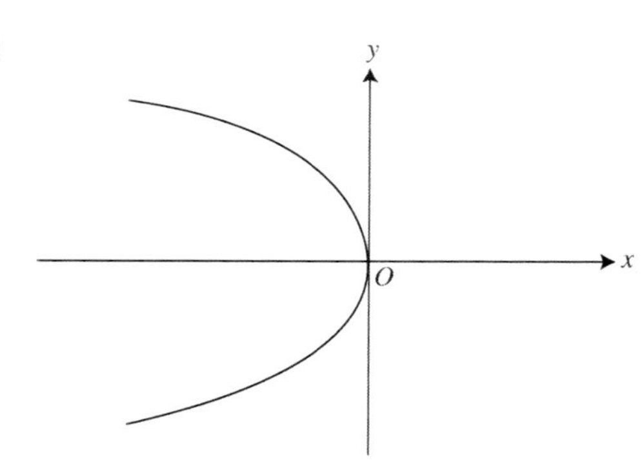

$y^2 = -4ax$

if $x = -a$, $y^2 = 4a^2 \Rightarrow y = \pm 2a$

if $x = -4a$, $y^2 = 16a^2 \Rightarrow y = \pm 4a$

if $x = 0$, $y = 0$.

(iii)

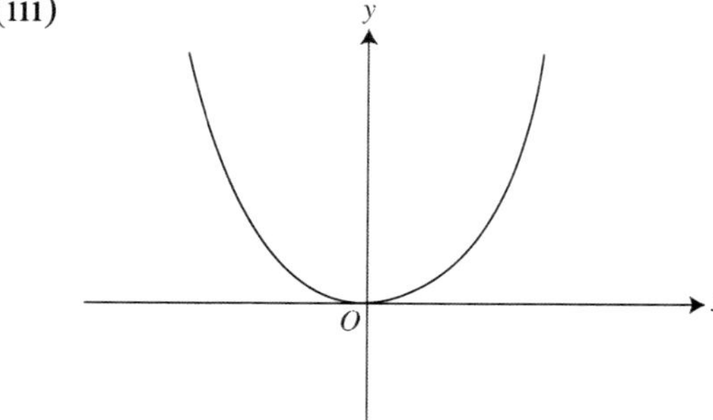

$x^2 = 4ay$

if $y = a$, $x^2 = 4a^2 \Rightarrow x = \pm 2a$

if $y = 4a$, $x^2 = 16a^2 \Rightarrow x = \pm 4a$

if $x = 0$, $y = 0$.

(iv)

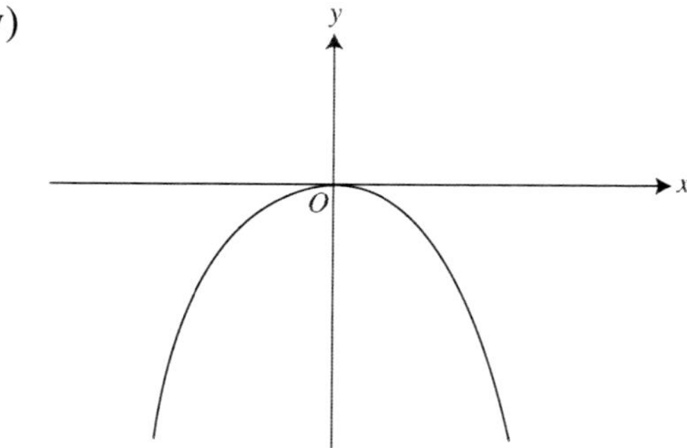

$x^2 = -4ay$

if $y = -a$, $x^2 = 4a^2 \Rightarrow x = \pm 2a$

if $y = -4a$, $x^2 = 16a^2 \Rightarrow x = \pm 4a$

if $x = 0$, $y = 0$.

---

2. (a)

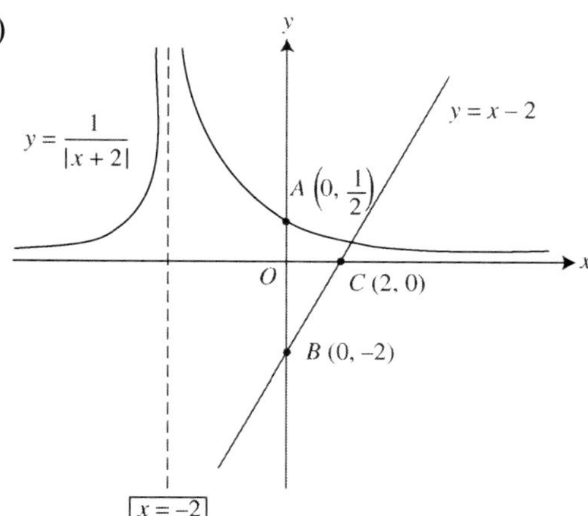

$y = x - 2$

if $x = 0$, $y = -2$, when $y = 0$, $x = 2$

$\boxed{y = 0}$ $\boxed{x = -2}$ asymtotes

(b) $x - 2 < \dfrac{1}{|x+2|}$ squaring up both sides $(x-2)^2 < \dfrac{1}{(x+2)^2}$

$$(x-2)^2\,(x+2)^2 < 1 \Rightarrow \big[(x-2)(x+2) - 1\big]\big[(x-2)(x+2) + 1\big] < 0$$

$$(x^2 - 4 - 1)(x^2 - 4 + 1) < 0 \Rightarrow (x^2 - 5)(x^2 - 3) < 0$$

$$-\sqrt{5} < x < -\sqrt{3} \qquad \sqrt{3} < x\sqrt{5}.$$

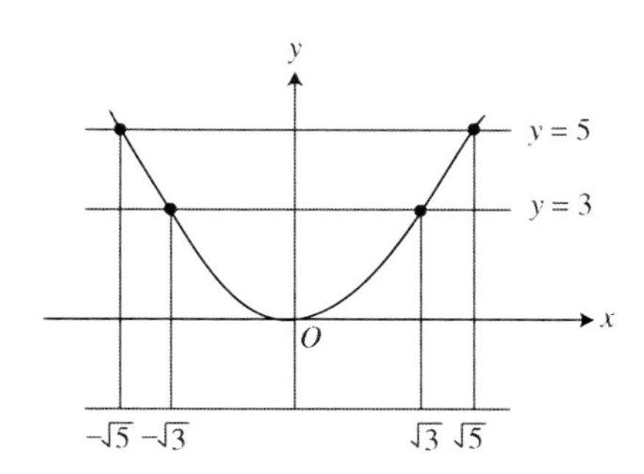

---

3. $\mathbf{AB} = \begin{pmatrix} 3 & -1 \\ -4 & 3 \end{pmatrix} \times \begin{pmatrix} \frac{3}{5} & \frac{1}{5} \\ \frac{4}{5} & \frac{3}{5} \end{pmatrix} = \begin{pmatrix} 3 \times \frac{3}{5} + -1 \times \frac{4}{5} & 3 \times \frac{1}{5} + -1 \times \frac{3}{5} \\ -4 \times \frac{3}{5} + 3 \times \frac{4}{5} & -4 \times \frac{1}{5} + 3 \times \frac{3}{5} \end{pmatrix}$

$= \begin{pmatrix} 1 & 0 \\ 0 & 1 \end{pmatrix}.$

$$\mathbf{BA} = \begin{pmatrix} \frac{3}{5} & \frac{1}{5} \\ \frac{4}{5} & \frac{3}{5} \end{pmatrix} \times \begin{pmatrix} 3 & -1 \\ -4 & 3 \end{pmatrix} = \begin{pmatrix} \frac{3}{5} \times 3 + \frac{1}{5} \times -4 & \frac{3}{5} \times -1 + \frac{1}{5} \times 3 \\ \frac{4}{5} \times 3 + \frac{3}{5} \times -4 & \frac{4}{5} \times -1 + \frac{3}{5} \times 3 \end{pmatrix}$$

$$= \begin{pmatrix} 1 & 0 \\ 0 & 1 \end{pmatrix}.$$

It is observed that $\mathbf{B} = \mathbf{A}^{-1}$, therefore $\mathbf{AB} = \mathbf{BA} = \mathbf{I}$ and shows that $\mathbf{A}$ and $\mathbf{B}$ are commutative. In general, however the product is non-commutative.

---

4. $z^2 - 4z + 8 = 0 \qquad z = \dfrac{4 \pm \sqrt{16 - 32}}{2} = 2 \pm i2$

$z_1 = 2 + i2 \qquad z_2 = 2 - i2$

$|z_1| = |2 + i2| = \sqrt{4 + 4} = 2\sqrt{2} \qquad |z_2| = \sqrt{2^2 + (-2)^2} = 2\sqrt{2}$

$\arg z_1 = \tan^{-1} \dfrac{2}{2} = \tan^{-1} = 45° = \dfrac{\pi}{4}$

$\arg z_2 = -\tan^{-1} \dfrac{2}{2} = -\tan^{-1} 1 = -45° = -\dfrac{\pi}{4}.$

---

5. Let the quadratic function be $f(x) = ax^2 + bx + c \;\; a > 0$ since it has a minimum point $(-2, 2)$. The discriminant $D = b^2 - 4ac < 0$ since the graph neither touches nor intersects the x-axis,

$f(0) = c = 5$

$$f(x) = a\left[x^2 + \frac{b}{a}x + \frac{c}{a}\right] = a\left[\left(x + \frac{b}{2a}\right)^2 - \frac{b^2}{4a^2} + \frac{c}{a}\right]$$

$$f\left(-\frac{b}{2a}\right) = a\left[-\frac{b^2}{4a^2} + \frac{c}{a}\right] = -\frac{b^2}{4a} + c \Rightarrow f(-2) = -\frac{b^2}{4a} + 5 = 2$$

$$\frac{b^2}{4a} = 3 \text{ but } -\frac{b}{2a} = -2 \Rightarrow b = 4a$$

$$b^2 = 12a \Rightarrow 16a^2 = 12a \Rightarrow 4a^2 = 3a \Rightarrow 4a = 3$$

$$a = \frac{3}{4} \text{ and } b = 4 \times \frac{3}{4} = 3, \qquad c = 5 \;\therefore f(x) = \frac{3}{4}x^2 + 3x + 5.$$

---

6. $\begin{pmatrix} 1 & 0 \\ 0 & -1 \end{pmatrix}\begin{pmatrix} -3 & 3 & 2 & -4 \\ -4 & -2 & -6 & -8 \end{pmatrix} = \begin{pmatrix} -3 & 3 & 2 & -4 \\ 4 & 2 & 6 & 8 \end{pmatrix}$

A′ B′ C′ D′ A B C D

---

7. (a) Let $x = \alpha$ and $x = \beta$ be the roots of the quadratic equation $ax^2 + bx + c = 0$.

$x = \alpha \Rightarrow x - \alpha = 0, \qquad x = \beta \Rightarrow x - \beta = 0$

$(x - \alpha)(x - \beta) = 0 = x^2 - \alpha x - \beta x + \alpha\beta = 0$

$= x^2 - (\alpha + \beta)x + \alpha\beta = 0 \qquad \ldots \textbf{(1)}$

where $S = \alpha + \beta$ and $P = \alpha\beta$

$ax^2 + bx + c = 0$ dividing each term by a where

$a \neq 0, \quad x^2 + \frac{b}{a}x + \frac{c}{a} = 0 \qquad \ldots \textbf{(2)}$

comparing equations (1) and (2) $\quad S = \alpha + \beta = -\frac{b}{a}$ and $P = \alpha\beta = \frac{c}{a}$.

(b) $ax^2 + bx + c = 0$, dividing each term by a $\quad x^2 + \frac{b}{a}x + \frac{c}{a} = 0$

look at the coefficient of x, halve it, square it and complete it

$$\left(x + \frac{b}{2a}\right)^2 - \frac{b^2}{4a^2} + \frac{c}{a} = 0$$

$$\left(x + \frac{b}{2a}\right)^2 = \frac{b^2 - 4ac}{4a^2}, \text{ square rooting both sides}$$

$$x + \frac{b}{2a} = \pm\frac{\sqrt{b^2 - 4ac}}{2a} \Rightarrow x = \frac{-b \pm \sqrt{b^2 - 4ac}}{2a}$$

$$\alpha + \beta = \frac{-b + \sqrt{b^2 - 4ac}}{2a} + \frac{-b - \sqrt{b^2 - 4ac}}{2a} = -\frac{b}{a} \Rightarrow \boxed{\alpha + \beta = -\frac{b}{a}}$$

$$\alpha\beta = \frac{(-b + \sqrt{b^2 - 4ac})}{2a}\,\frac{(-b - \sqrt{b^2 - 4ac})}{2a} = \left(-\frac{b}{2a}\right)^2 - \left(\frac{\sqrt{b^2 - 4ac}}{2a}\right)^2$$

$$\boxed{\alpha\beta = \frac{c}{a}}$$

---

8. (i) $1 + 2 + 3 + \cdots + \cdots + (n-2) + (n-1) + n = \sum_{r=1}^{n} r \qquad \ldots \mathbf{(1)}$

this can be written

$$n + (n-1) + (n-2) + \cdots + 3 + 2 + 1 = \sum_{r=1}^{n} r \qquad \ldots \mathbf{(2)}$$

adding equations (1) and (2)

$$(1+n) + (1+n) + (1+n) + \cdots + (n+1) + (n+1) + (n+1) = 2\sum_{r=1}^{n} r$$

$$\therefore 2\sum_{r=1}^{n} r = n(n+1) \Rightarrow \boxed{\sum_{r=1}^{n} r = \frac{n(n+1)}{2}}$$

(ii)

$$n^2 - \cancel{(n-1)}^2 = \qquad n^2 - n^2 + 2n - 1 = 2n - 1$$

$$\cancel{(n-1)}^2 - (n-1)^2 = n^2 - 2n + 1 - n^2 + 4n - 4 = 2(n-1) - 1$$

$$(n-2)^2 - (n-3)^2 = n^2 - 4n + 4 - n^2 + 6n - 9 = 2(n-2) - 1$$

$$(n-3)^2 - (n-4)^2 = n^2 - 6n + 9 - n^2 + 8n - 16 = 2(n-3) - 1$$

$$\vdots$$

$$2^2 - \cancel{1^2} = \qquad = 2 \times 2 - 1$$

$$\cancel{1^2} - 0^2 = \qquad = 2 \times 1 - 1$$

adding the left hand sides and the right hand sides, we have

$$n^2 = 2\sum_{r=1}^{n} r - n \qquad \therefore \boxed{\sum_{r=1}^{n} r = \frac{n(n+1)}{2}}$$

---

9. (a) $z^2 + z + 1 = 0 \Rightarrow z = \dfrac{-1 \pm \sqrt{1-4}}{2}$ or $-\dfrac{1}{2} \pm i\dfrac{\sqrt{3}}{2}$

(b)

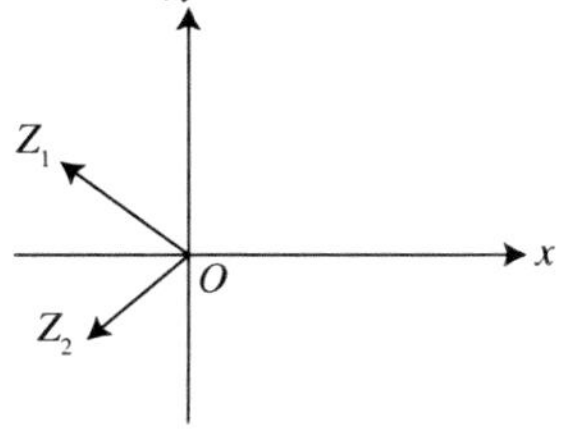

$$z_1 = -\frac{1}{2} + i\frac{\sqrt{3}}{2} \text{ or } z_2 = -\frac{1}{2} - i\frac{\sqrt{3}}{2}.$$

(c) (i) $S = \alpha + \beta = z_1 + z_2 = -1$

(ii) $P = \alpha\beta = z_1 z_2 = 1.$

(iii) $\frac{z_1}{z_2} = \frac{-\frac{1}{2} + i\frac{\sqrt{3}}{2}}{-\frac{1}{2} - i\frac{\sqrt{3}}{2}} = \frac{-1 + i\sqrt{3}}{-1 - i\sqrt{3}} \times \frac{-1 + i\sqrt{3}}{-1 + i\sqrt{3}}$

$$\mathrm{W} = \frac{z_1}{z_2} = \frac{(-1 + i\sqrt{3})^2(-1 + i\sqrt{3})}{(-1)^2 - (i\sqrt{3})^2} = \frac{-1 - 2i\sqrt{3} - 3}{4} = -\frac{1}{2} - \frac{1}{2}\sqrt{3}i$$

(iv) $\frac{z_2}{z_1} = \frac{-\frac{1}{2} - i\frac{\sqrt{3}}{2}}{-\frac{1}{2} + i\frac{\sqrt{3}}{2}} = \frac{-1 - i\sqrt{3}}{-1 + i\sqrt{3}} \times \frac{-1 - i\sqrt{3}}{-1 - i\sqrt{3}} = \frac{(-1 - i\sqrt{3})^2}{(-1^2) - (i\sqrt{3})^2}$

$$= -\frac{1}{2} + \frac{\sqrt{3}}{2}i = W^*$$

therefore $\frac{z_1}{z_2}$ and $\frac{z_2}{z_1}$ are conjugates.

Examination Test Papers.

GCE Advanced Level.

Further Pure Mathematics FP1

Test Paper 2

Time: 1 hour and 30 minutes.

Instructions and Information

Candidates may use any calculator allowed by the regulations of their Examination Board.

Full marks are awarded for correct answers to ALL questions.

This paper has nine questions.

You can start working with any question and you must label clearly all parts.

1. One root of the quadratic equation $z^2 - 2z + 2 = 0$ is $z = 1 + i$, find the other root. If α and β are these two roots show that $\alpha + \beta = 2$ and $\alpha\beta = 2$. **(4)**

---

2. The equation $f(x) = 3^x + 3x - 7 = 0$ has an approximate value of $x = 1.25$. Use Newton-Raphson method twice to show that an improved approximation is found to be $x = 1.15183$, by using the formula twice. **(8)**

---

3. If α, β and γ are the roots of the cubic equation $x^3 - 5x^2 + 11x + 5 = 0$, determine the sum and the product of these roots. **(5)**

---

4. The locus of a point $P(x, y)$ which moves so that it is equidistant from the fixed point $S(a, 0)$ and the fixed line $x = -a$, is a parabola.
If $PN = PS$, show that $y^2 = 4ax$.

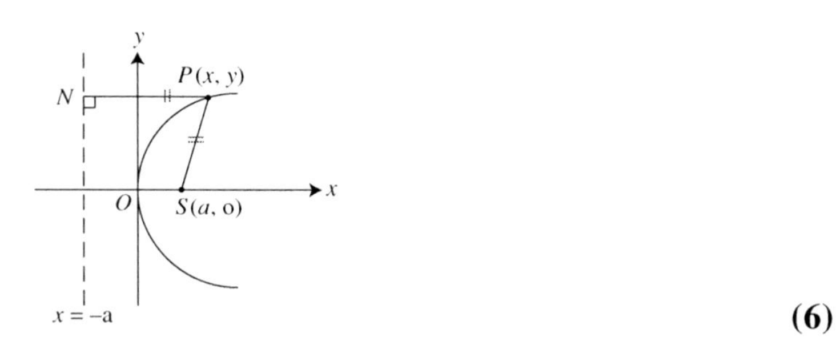

(6)

---

5. Given the 2×2 matrix $\mathbf{A}$

$$\mathbf{A} = \begin{pmatrix} a_{11} & a_{12} \\ a_{21} & a_{22} \end{pmatrix}$$

determine the inverse or reciprocal matrix, $\mathbf{A}^{-1}$, explaing the intermediate steps.

Explain $\mathbf{A}\mathbf{A}^{-1} = \mathbf{A}^{-1}\mathbf{A} = \mathbf{I}$. **(8)**

---

6. A parabola has equation $y^2 = 4ax$, $a > 0$.
The point $P(ap^2, 2ap)$ lies on the parabola. Find the equations of the tangent and normal at P.

Hence show that the coordinates of the points of intersection with the axes are:

$C\,(0, ap)$, $D(-ap^2, 0)$, $A(0, 2ap + ap^3)$ and $B(2a + ap^2, 0)$.

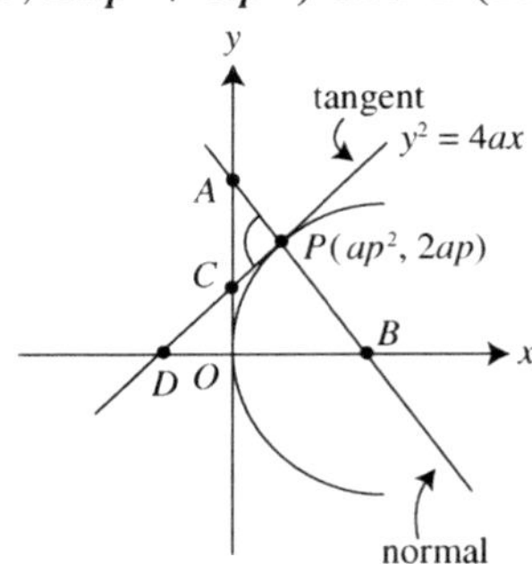

(8)

---

7. If α and β are the roots of the quadratic equation $ax^2 + bx + c = 0$,

find the values of (i) $\alpha^3 + \beta^3$ and (ii) $\alpha^3 - \beta^3$ in terms of a and b. **(8)**

Hence show that the numerical values of (i) and (ii) are -18 and 8 $\sqrt{5}$

respectively if $b = 3$ and $a = c = 1$. **(4)**

---

8. Prove that $\sum_{r=1}^{n} r^2 = \frac{n}{6}(n+1)(2n+1)$ using the difference of squares method.

Hence evaluate **(10)**

$$\sum_{r=1}^{10} r(r+1).$$ **(2)**

---

9. A right angled triangle T has vertices A (2, 2) , B (4, 2) and C (4, 6).

When T is transformed by the matrix P = $\begin{pmatrix} 0 & -1 \\ 1 & 0 \end{pmatrix}$, the image is T'.

(a) Find the coordinates of the vertices of T'. **(2)**

(b) Describe fully the transformation represented by P. **(2)**

Draw the object and image. **(2)**

(c) The matrices $\mathbf{M} = \begin{pmatrix} 2 & -3 \\ 4 & -5 \end{pmatrix}$ and $\mathbf{N} = \begin{pmatrix} 1 & 2 \\ 3 & 4 \end{pmatrix}$

represent two transformations.

When T' is transformed by the image **MN** the image is T". Find **MN**. **(2)**

(d) Find the determinate of MN. **(2)**

(e) Using your answer to part (d) find the area of T". **(2)**

TOTAL FOR PAPER: 75 MARKS

Examination Test Papers.

GCE Advanced Level.

Test Paper 2 Solutions

Further Pure Mathematics FP1

1. $z^2 - 2z + 2 = 0$ observe that the coefficient of the quadratic function are real, $1, -2, 2$ therefore if $z = 1 + i$, the other root is $z^* = 1 - i$.

$\alpha + \beta = 1 + i + 1 - i = 2 \qquad \alpha\beta = (1 - i)(1 + i) = 1^2 - i^2 = 1 + 1 = 2.$

---

2. $x_{n+1} = x_n - \dfrac{f(x_n)}{f'(x_n)}$ the Newton-Raphson formula.

$x_n = 1.25 \qquad \therefore f(x_n) = f(1.25) = 3^{1.25} + 3 \times 1.25 - 7 = 0.698222038$

$f(x) = 3^x + 3x - 7, \qquad f'(x) = 3^x \ln 3 + 3 \qquad f'(1.25) = 3^{1.25} \ln 3 + 3 = 7.33756525$

$x_{n+1} = 1.25 - \dfrac{0.698222038}{7.33756525} = 1.154842817$

$f(1.154842817) = 3^{1.154842817} + 3(1.154842817) - 7 = 0.0208420777$

$f'(1.154842817) = 3^{1.154842817} \ln 3 + 3 = 6.907009853$

$x_{n+2} = 1.154842817 - \dfrac{0.0208420777}{6.907009853} = 1.151825292 = 1.15183$ to 6 s.f.

---

3. $(x - \alpha)(x - \beta)(x - \gamma) = \left[x^2 - (\alpha + \beta)x + \alpha\beta\right](x - \gamma)$

$$= x^3 - (\alpha + \beta + \gamma)x^2 + (\alpha\beta + \alpha\gamma + \beta\gamma)x - \alpha\beta\gamma = 0$$

comparing the coefficients of this cubic equation with $x^3 - 5x^2 + 11x + 5 = 0$

$\therefore \alpha + \beta + \gamma = 5$ and $\alpha\beta\gamma = -5$.

4.

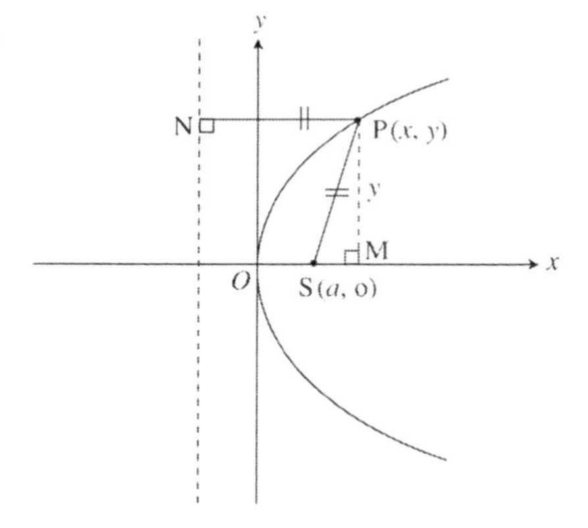

PS = PN. The right angle triangle SPM, we have $PS^2 = y^2 + (x - a)^2$

$PN = x + a \qquad \therefore PS^2 = PN^2$

$y^2 + (x - a)^2 = (x + a)^2$

$y^2 = (x + a)^2 - (x - a)^2 = \left[(x + a) - (x - a)\right]\left[(x + a) + (x - a)\right]$

$$= 2a \times 2x \qquad \boxed{y^2 = 4ax}$$

---

5. $\mathbf{A} = \begin{pmatrix} a_{11} & a_{12} \\ a_{21} & a_{22} \end{pmatrix}$

The minor of the element a_{11} is found by deleting the row containing a_{11} and deleting the column containing a_{11}.

thus the minor of a_{11} is a_{22} $\quad \begin{matrix} \not{a}_{11} & \not{a}_{12} \\ \not{a}_{21} & a_{22} \end{matrix}$

The minor of a_{12} is a_{21} since $\quad \begin{matrix} \not{a}_{11} & \not{a}_{12} \\ a_{21} & \not{a}_{22} \end{matrix}$

and the minor of a_{21} is a_{12} and that of a_{22} is a_{11}.

$\therefore$ The minors of $\mathbf{A}$ are $\begin{pmatrix} a_{22} & a_{21} \\ a_{12} & a_{11} \end{pmatrix}$.

$\mathbf{A}^*$ denotes the cofactors of the elements of the matrix $\mathbf{A}$.

The cofactors of $\mathbf{A}$, $\mathbf{A}^*$, are found by writing plus and minus alternatively starting with a plus at the upper left.

$$\mathbf{A}^* = \begin{pmatrix} a_{22} & -a_{21} \\ -a_{12} & a_{11} \end{pmatrix}$$

$\mathbf{A}^{*\mathrm{T}}$, the transpose of $\mathbf{A}^*$, that is, the columns are written as rows and the rows as columns.

$$\mathbf{A}^{*\mathrm{T}} = \begin{pmatrix} a_{22} & -a_{12} \\ -a_{21} & a_{11} \end{pmatrix} = \text{the adjoint matrix of } \mathbf{A}.$$

Finally $\mathbf{A}^{-1} = \dfrac{\mathbf{A}^{*\mathrm{T}}}{|\mathbf{A}|}$ where $|\mathbf{A}|$ is the determinant of $\mathbf{A}$.

Note: $\mathbf{AA}^{-1} = \mathbf{A}^{-1}\mathbf{A} = \mathbf{I} = \begin{pmatrix} 1 & 0 \\ 0 & 1 \end{pmatrix}$ post or pre-multiplication of the matrices and their inverses is commutative.

---

6. $y^2 = 4ax \Rightarrow 2y\dfrac{\mathrm{d}y}{\mathrm{d}x} = 4a \Rightarrow \dfrac{\mathrm{d}y}{\mathrm{d}x} = \dfrac{2a}{y} = \dfrac{2a}{2ap} = \dfrac{1}{p}$

$y = mx + c$ the equation of the tangent $m = \dfrac{\mathrm{d}y}{\mathrm{d}x} = \dfrac{1}{p}$ = the gradient of the tangent

$\therefore y = \dfrac{1}{p}x + c \Rightarrow 2ap = \dfrac{1}{p}ap^2 + c \qquad c = 2ap - ap = ap$

$\therefore y = \dfrac{1}{p}x + ap \qquad yp = x + ap^2 \qquad \ldots(\mathbf{1})$

The gradient of the normal is $-p \Rightarrow y = -px + c$

$2ap = -pap^2 + c \Rightarrow c = 2ap + ap^3 \qquad \therefore \boxed{y = -px + 2ap + ap^3} \qquad \ldots(\mathbf{2})$

when $x = 0$ in(1) $y = ap$ $C(0, ap)$ when $y = 0$, $x = -ap^2$ $D(-ap^2, 0)$,

when $x = 0$ in(2) $y = 2ap + ap^3$ $A(0, 2ap + ap^3)$,

when $y = 0$ $x = 2a + ap^2$, $B(2a + ap^2, 0)$.

---

7. (i) $\alpha^3 + \beta^3 = (\alpha + \beta)^3 - 3\alpha^2\beta - 3\alpha\beta^2$ but $\alpha + \beta = -\frac{b}{a}$ and $\alpha\beta = \frac{c}{a}$

$$\alpha^3 + \beta^3 = \left(-\frac{b}{a}\right)^3 - 3\alpha\beta(\alpha + \beta) = -\frac{b^3}{a^3} - 3\frac{c}{a}\left(-\frac{b}{a}\right)$$

$$= -\frac{b^3}{a^3} + \frac{3bc}{a^2} = \frac{3abc - b^3}{a^3} = \frac{3 \times 1 \times 3 \times 1 - 3^3}{1^3} = -18$$

(ii) $\alpha^3 - \beta^3 = (\alpha - \beta)(\alpha^2 + \alpha\beta + \beta^2)$

$$= (\alpha - \beta)\left[(\alpha + \beta)^2 - 2\alpha\beta + \alpha\beta\right] = (\alpha - \beta)\left[(\alpha + \beta)^2 - \alpha\beta\right]$$

$$= \left(-\frac{b}{2a} + \frac{\sqrt{b^2 - 4ac}}{2a} + \frac{b}{2a} + \frac{\sqrt{b^2 - 4ac}}{2a}\right)\left[\left(-\frac{b}{a}\right)^2 - \frac{c}{a}\right]$$

$$= \frac{\sqrt{b^2 - 4ac}}{a}\left(\frac{b^2}{a^2} - \frac{c}{a}\right) = \frac{\sqrt{b^2 - 4ac}\,(b^2 - ac)}{a^3}$$

$$= \frac{\sqrt{b^2 - 4ac}}{a^3}\left(b^2 - ac\right) = \frac{\sqrt{9 - 4}}{1}\left(9 - 1\right) = 8\sqrt{5}.$$

---

8. $n^3 - (\cancel{n-1})^3 = n^3 - n^3 + 3n^2 - 3n + 1 \qquad = 3n^2 - 3n + 1$

$(\cancel{n-1})^3 - (n-2)^2 = n^3 - 3n^2 + 3n - 1 - n^3 + 6n^2 - 12n + 8 = 3(n-1)^2 - 3(n-1) + 1$

$(n-2)^3 - (n-3)^3 = \qquad = 3(n-2)^2 - 3(n-2) + 1$

$\vdots$

$\qquad = 3 \times 1^2 - 3 \times 1 + 1$

$\dfrac{\cancel{1^3} - 0^3}{n^3} = \qquad = 3\displaystyle\sum_{r=1}^{n} r^2 - 3\sum_{r=1}^{n} r + n \qquad \ldots(\mathbf{1})$

where $(a+b)^3 = a^3 + 3a^2b + 3ab^2 + b^3$

$$(n-1)^3 = n^3 - 3n^2 + 3n - 1$$

$$(n-2)^3 = n^3 - 3 \times 2n^2 + 3(-2)^2 n - 8 = n^3 - 6n^2 + 12n - 8$$

$$\begin{aligned}(n-1)^3 - (n-2)^3 &= n^3 - 3n^2 + 3n - 1 - n^3 + 6n^2 - 12n + 8 \\ &= 3n^2 - 9n + 7 = 3(n-1)^2 - 3(n-1) + 1 \\ &= 3n^2 - 6n + 3 - 3n + 3 + 1 \\ &= 3n^2 - 9n + 7\end{aligned}$$

from (1)

$$n^3 = 3\sum_{r=1}^{n} r^2 - 3\sum_{r=1}^{n} r + n$$

$$3\sum_{r=1}^{n} r^2 = n^3 + \frac{3(n+1)n}{2} - n = \frac{2n^2 + 3n(n+1) - 2n}{2}$$

$$\sum_{r=1}^{n} r^2 = \frac{n}{6}\left(2n^2 + 3n + 1\right) = \frac{n}{6}\left(n+1\right)(2n+1)$$

$$\therefore \boxed{\sum_{r=1}^{n} r^2 = \frac{n(n+1)(2n+1)}{6}}$$

$$\sum_{r=1}^{10} r(r+1) = \sum_{r=1}^{10} r^2 + \sum_{r=1}^{10} r = \frac{10 \times 11 \times 21}{6} + \frac{10 \times 11}{2} = 440.$$

---

9. (a) $T' = \begin{pmatrix} 0 & -1 \\ 1 & 0 \end{pmatrix}\begin{pmatrix} 2 & 4 & 4 \\ 2 & 2 & 6 \end{pmatrix} = \begin{pmatrix} -2 & -2 & -6 \\ 2 & 4 & 4 \end{pmatrix}$

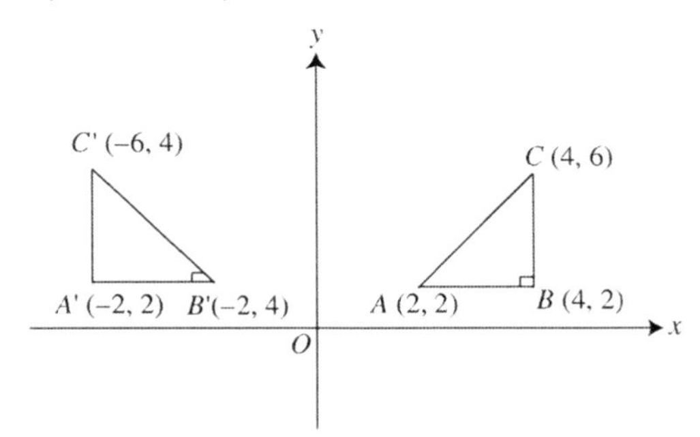

(b) A rotation of 90° about the origin in an anticlockwise direction.

(c) $\mathbf{MN} = \begin{pmatrix} 2 & -3 \\ 4 & -5 \end{pmatrix}\begin{pmatrix} 1 & 2 \\ 3 & 4 \end{pmatrix} = \begin{pmatrix} -7 & -8 \\ -11 & -12 \end{pmatrix}$.

(d) $|\mathbf{MN}| = |84 - 88| = 4$.

(e) Area of image = area of object $\times 4 = 4 \times 4 = 16$ s.u.

Examination Test Papers.

GCE Advanced Level.

Further Pure Mathematics FP1

Test Paper 3

Time: 1 hour and 30 minutes.

Instructions and Information

Candidates may use any calculator allowed by the regulations of their Examination Board.

Full marks are awarded for correct answers to ALL questions.

This paper has nine questions.

You can start working with any question and you must label clearly all parts.

1. One root of the cubic equation

$z^3 - 5z^2 + 8z - 6 = 0$ is $z = 1 - i$ find the other two roots. **(6)**

---

2. (a) Use the formula $\sum_{r=1}^{n} r = \frac{n(n+1)}{2}$ to show that $\sum_{r=1}^{n}(2r-1) = n^2$. **(2)**

(b) Prove by the method of mathematiical induction, that, for $n \in \mathbb{Z}^+$,

$\sum_{r=1}^{n}(2r-1) = n^2$. **(6)**

---

3. An approximate root for $f(x) = 3^x + 2x + 4$ is given as -2.03125.

Use Newton-Raphson method to find an improved root to 5 decimal places.

Check your result. **(6)**

---

4. Show by the methods of mathematical induction that $\begin{pmatrix} 1 & 2 \\ 0 & 1 \end{pmatrix}^n = \begin{pmatrix} 1 & 2n \\ 0 & 1 \end{pmatrix}$

for all positive integers. **(8)**

---

5. A square ABCD has the coordinate matrix $\begin{pmatrix} 2 & 3 & 3 & 2 \\ 1 & 1 & 2 & 2 \end{pmatrix}$.

A B C D

Determine the coordinate matrix of the reflected A′B′C′D′ square in the y-axis and A"B"C"D" square in the x-axis. **(4)**

---

6. (a) Find the roots of the equation $z^2 + z + 1 = 0$, giving your answers in the form $a + ib$, z_1 and z_2. **(2)**

(b) show these roots z_1 and z_2 on an Argand diagram. **(2)**

(c) Find (i) the sum of the roots (ii) the product of the roots (iii) $\frac{z_1}{z_2}$

(iv) $\frac{z_2}{z_1}$ and show that these are conjugates. **(6)**

---

7. Determine the equations of the parabolas with parametric equations:

(i) $x = 5t^2$, $y = 3t$ (ii) $x = -3p^2$, $y = 5p$ (iii) $x = 3t^2$, $y = 6t$

(iv) $x = -2t$, $y = -3t^2$. Sketch these parabolas.

Hence state the coordinates of the foci and the equations of the directrices. **(12)**

---

8. (a) Use interval bisection to find the positive root of $x = \sqrt{13}$ to

3 significant figures. Use five approximations. **(5)**

(b) Use Newton-Raphson method to find an improved root, use the value found in the fifth approximation to show that $x = 3.605506796$. **(6)**

---

9. (a) Using the formulae for $\sum_{r=1}^{n} r$, $\sum_{r=1}^{n} r^2$

and $\sum_{r=1}^{n} r^3$, show that $\sum_{r=1}^{n} r(r+1)(r+2) = \frac{n(n+1)(n+2)(n+3)}{4}$.

(b) Hence evaluate $\sum_{r=11}^{20} r(r+1)(r+2)$. **(10)**

TOTAL FOR PAPER: 75 MARKS

Examination Test Papers.

GCE Advanced Level.

Test Paper 3 Solutions

Further Pure Mathematics FP1

1. $z^3 - 5z^2 + 8z - 6 = 0.$

The coefficients of the cubic equation are real, if $z = 1 + i$, the other complex root is $z = 1 - i$, which is the conjugate, z^*, of $z = 1 + i$.

$(z - 1 - i)(z - 1 + i) = (z - 1)^2 - i^2 = z^2 - 2z + 1 + 1 = z^2 - 2z + 2$

$$\begin{array}{r|l} & z - 3 \\ \hline z^2 - 2z + 2 & z^3 - 5z^2 + 8z - 6 \\ & \underline{z^3 - 2z^2 + 2z} \\ & \quad -3z^2 + 6z - 6 \\ & \quad \underline{-3z^2 + 6z - 6} \\ & \qquad\quad 0 \end{array}$$

$\therefore z^3 - 5z^2 + 8z - 6 = (z - 3)(z^2 - 2z + 2) = 0$

and the third root is $z = 3$.

---

2. (a) $\sum_{r=1}^{n}(2r - 1) = 2\sum_{r=1}^{n} r - \sum_{r=1}^{n} 1 = 2 \times \dfrac{n(n+1)}{2} - n = n^2 + n - n = n^2.$

(b) Method of mathematical induction

$\sum_{r=1}^{n}(2r - 1) = 1 + 3 + 5 + \cdots + 2n - 1 = n^2 \qquad \ldots (1)$

if $n = 1 \qquad 1 = 1^2,$ it is true from (1)

if $n = 2, \qquad 1 + 3 = 2^2,$ it is also true from (1)

if $n = k, \qquad 1 + 3 + 5 + \ldots 2k - 1 = k^2, \qquad \ldots (2),$ it would be true.

Adding the $(k + 1)^{\text{th}}$ term to both sides of (2)

$1 + 3 + 5 + \cdots + 2k - 1 + 2(k + 1) - 1 = k^2 + 2(k + 1) - 1$

$= k^2 + 2k + 1 = (k + 1)^2$

Therefore it is true for $n \qquad \therefore \sum_{r=1}^{n}(2r - 1) = n^2.$

---

3. $f(x) = 3^x + 2x + 4$

$f(-2.03125) = 3^{-2.03125} + 2(-2.03125) + 4 = 0.044861223$

$f'(x) = 3^x \ln 3 + 2$

$f'(-2.03125) = 3^{-2.03125} \ln 3 + 2 = 2.117948359$

$x_{n+1} = x_n - \dfrac{f(x_n)}{f'(x_n)} = -2.03125 - \dfrac{0.044861223}{2.117948359} = 2.053554505$

$= -2.05355$ to 5 d.p.

Check $f(-2.053554505) = 3^{-2.053554505} + 2(-2.053554505) + 4 = -0.002346595.$

---

4. $\begin{pmatrix} 1 & 2 \\ 0 & 1 \end{pmatrix}^n = \begin{pmatrix} 1 & 2n \\ 0 & 1 \end{pmatrix}^n \quad \ldots (1)$

for $n = 1$ from (1)

$\begin{pmatrix} 1 & 2 \\ 0 & 1 \end{pmatrix} = \begin{pmatrix} 1 & 2 \\ 0 & 1 \end{pmatrix}$ it is true

for $n = 2$

$\begin{pmatrix} 1 & 2 \\ 0 & 1 \end{pmatrix}^2 = \begin{pmatrix} 1 & 2 \\ 0 & 1 \end{pmatrix}\begin{pmatrix} 1 & 2 \\ 0 & 1 \end{pmatrix} = \begin{pmatrix} 1 & 4 \\ 0 & 1 \end{pmatrix} = \begin{pmatrix} 1 & 2 \times 2 \\ 0 & 1 \end{pmatrix}$ it is also true

for $n = k$

$\begin{pmatrix} 1 & 2 \\ 0 & 1 \end{pmatrix}^k = \begin{pmatrix} 1 & 2k \\ 0 & 1 \end{pmatrix}$ it would be true

$\begin{pmatrix} 1 & 2 \\ 0 & 1 \end{pmatrix}^{k+1} = \begin{pmatrix} 1 & 2 \\ 0 & 1 \end{pmatrix}^k \begin{pmatrix} 1 & 2 \\ 0 & 1 \end{pmatrix} = \begin{pmatrix} 1 & 2k \\ 0 & 1 \end{pmatrix}\begin{pmatrix} 1 & 2 \\ 0 & 1 \end{pmatrix}$

$= \begin{pmatrix} 1+0 & 2+2k \\ 0+0 & 0+1 \end{pmatrix} = \begin{pmatrix} 1 & 2(k+1) \\ 0 & 1 \end{pmatrix}$

$\therefore \begin{pmatrix} 1 & 2 \\ 0 & 1 \end{pmatrix}^n = \begin{pmatrix} 1 & 2n \\ 0 & 1 \end{pmatrix}$ it is true for n.

---

5. $\begin{pmatrix} -1 & 0 \\ 0 & 1 \end{pmatrix} \overset{\text{A B C D}}{\begin{pmatrix} 2 & 3 & 3 & 2 \\ 1 & 1 & 2 & 2 \end{pmatrix}} = \underset{\text{A}' \;\; \text{B}' \;\; \text{C}' \;\; \text{D}'}{\begin{pmatrix} -2 & -3 & -3 & -2 \\ 1 & 1 & 2 & 2 \end{pmatrix}}$

$\begin{pmatrix} 1 & 0 \\ 0 & -1 \end{pmatrix}\begin{pmatrix} 2 & 3 & 3 & 2 \\ 1 & 1 & 2 & 2 \end{pmatrix} = \underset{\text{A}'' \;\; \text{B}'' \;\; \text{C}'' \;\; \text{D}''}{\begin{pmatrix} 2 & 3 & 3 & 2 \\ -1 & -1 & -2 & -2 \end{pmatrix}}$

---

6. (a) $z^2 + z + 1 = 0$

$$z = \frac{-1 \pm \sqrt{1-4}}{2} \Rightarrow z_1 = -\frac{1}{2} + i\frac{\sqrt{3}}{2},\ z_2 = -\frac{1}{2} - i\frac{\sqrt{3}}{2}.$$

(b)

(c) (i) the sum of the roots, $z_1 + z_2 = -1$

(ii) the product of the roots, $z_1 z_2 = 1$

(iii) $$\frac{z_1}{z_2} = \frac{-\frac{1}{2} + i\frac{\sqrt{3}}{2}}{-\frac{1}{2} - i\frac{\sqrt{3}}{2}} = \frac{-1 + i\sqrt{3}}{-1 - i\sqrt{3}} \times \frac{-1 + i\sqrt{3}}{-1 - i\sqrt{3}} = \frac{-2 - 2i\sqrt{3}}{4}$$

$$= -\frac{1}{2} - \frac{1}{2}\sqrt{3}i = W$$

(iv) $$\frac{z_2}{z_1} = \frac{-\frac{1}{2} + i\frac{\sqrt{3}}{2}}{-\frac{1}{2} + i\frac{\sqrt{3}}{2}} = \frac{-1 - i\sqrt{3}}{-1 + i\sqrt{3}} \times \frac{-1 - i\sqrt{3}}{-1 - i\sqrt{3}} = \frac{(-1 - i\sqrt{3})^2}{(-1)^2 - i^2(\sqrt{3})^2}$$

$$= \frac{1 - 3 + 2i\sqrt{3}}{4} = -\frac{1}{2} + \frac{\sqrt{3}}{2}i = W^*$$

(iii) & (iv) are conjugates.

---

7. (i) $x = 5t^2$, $y = 3t$ eliminating the parameters, between these two equations, we have

$x = 5t^2 \quad \ldots (1) \quad \Rightarrow \quad t^2 = \dfrac{x}{5}$

$y = 3t \quad \ldots (2) \quad \Rightarrow \quad t = \dfrac{y}{3}$

$\dfrac{x}{5} = \left(\dfrac{y}{3}\right)^2 \quad \Rightarrow \quad \boxed{y^2 = \dfrac{9}{5}x}$

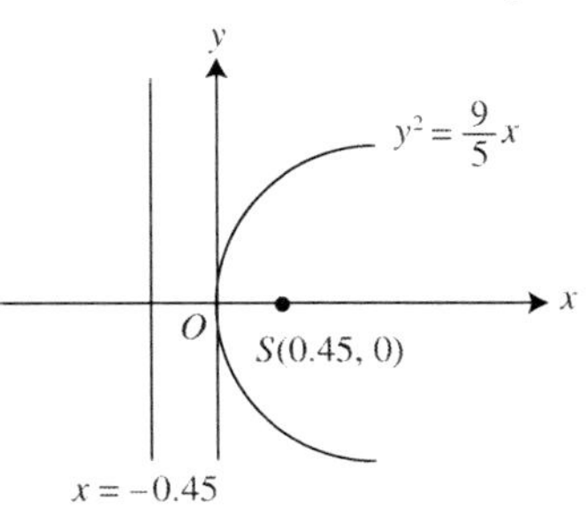

$$y^2 = \frac{9}{5}x = 4ax \quad \Rightarrow \quad 4a = \frac{9}{5} \quad \Rightarrow \quad a = \frac{9}{20} = 0.45$$

$\boxed{S(0.45, 0)}$ focus $\boxed{x = -0.45}$ directrix

(ii) $x = -3p^2 \quad \ldots (1)$

$y = 5p \quad \ldots (2)$

From (2) $p = \dfrac{y}{5}$, substitute in (1) $\quad x = -3\left(\dfrac{y}{5}\right)^2$

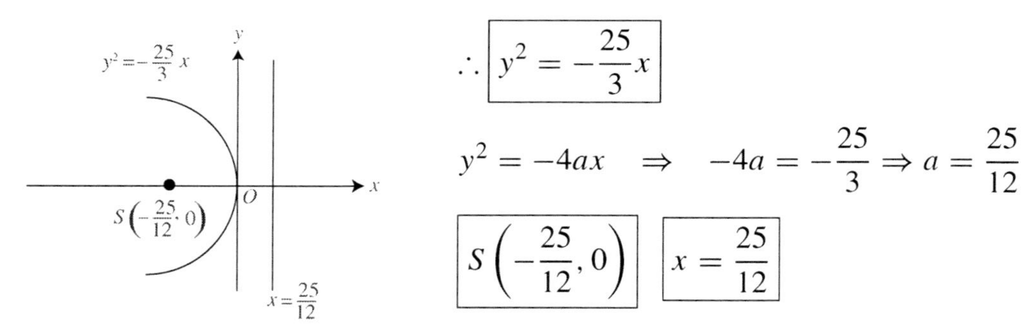

$\therefore \boxed{y^2 = -\dfrac{25}{3}x}$

$y^2 = -4ax \quad \Rightarrow \quad -4a = -\dfrac{25}{3} \Rightarrow a = \dfrac{25}{12}$

$\boxed{S\left(-\dfrac{25}{12}, 0\right)} \quad \boxed{x = \dfrac{25}{12}}$

(iii) $x = 3t^2 \quad \ldots (1)$

$y = 6t \quad \ldots (2)$

From (2) $\quad t = \dfrac{y}{6} \quad$ and substituting in (1)

$x = 3\left(\dfrac{y}{6}\right)^2 \quad \Rightarrow \quad 3y^2 = 36x \quad \Rightarrow \quad \boxed{y^2 = 12x}$

$y^2 = 12x = 4ax \quad \Rightarrow \quad \boxed{a = 3} \quad \boxed{S(3, 0)} \quad \boxed{x = -3}$ directrix

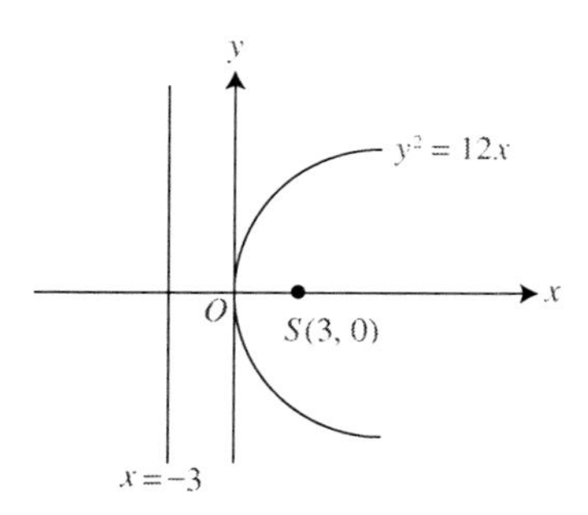

(iv) $x = -2t \quad \ldots (1)$

$y = -3t^2 \quad \ldots (2)$

From (1) $\quad t = -\dfrac{x}{2}$ and substituting in (2) $\quad y = -3\left(-\dfrac{x}{2}\right)^2$

$y = -\dfrac{3}{4}x^2$

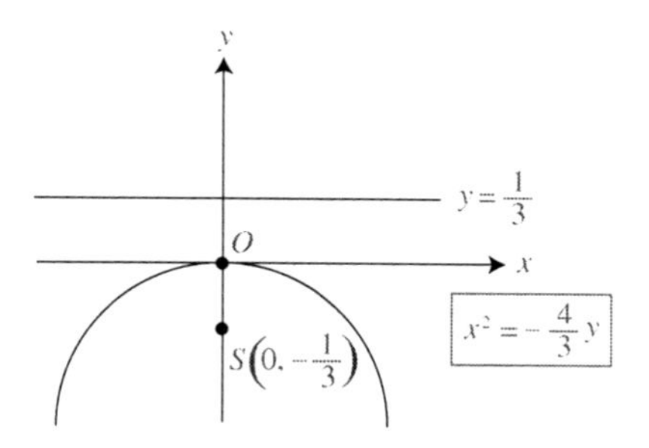

$$x^2 = -\frac{4}{3}y = -4ay \quad \Rightarrow \quad -4a = -\frac{4}{3} \quad \Rightarrow \quad a = \frac{1}{3}, \boxed{y = \frac{1}{3}} \quad \boxed{S\left(0, -\frac{1}{3}\right)}$$

---

8. (a) $x = \sqrt{13} \quad \Rightarrow \quad x^2 - 13 = 0 \quad \Rightarrow \quad \mathrm{f}(x) = x^2 - 13.$

Try values of x such that there is a change of sign.

Let $x = 3$, $\mathrm{f}(3) = 9 - 13 = -4$

let $x = 4$, $\mathrm{f}(4) = 16 - 13 = 3$.

Therefore there is a change of sign in the interval [3, 4] .

Using the interval bisection, that is, the mid-point, $\frac{3+4}{2} = 3.5$

$\mathrm{f}(3.5) = 3.5^2 - 13 = 12.25 - 13 = -0.75$ to get a positive answer

$\frac{3.5+4}{2} = \frac{7.5}{2} = 3.75$ the mid-point second approximation

$\mathrm{f}(3.75) = 3.75^2 - 13 = 1.0625$

in order to get a negative answer $\frac{3.75 + 3.50}{2} = 3.625$

the mid point, third approximation

$\mathrm{f}(3.625) = 3.625^2 - 13 = 0.140625$

in order to get a negative answer $\frac{3.625 + 3.50}{2} = 3.5625$ fourth approximation

$\mathrm{f}(3.5625) = 3.5625^2 - 13 = -0.30859375$ fifth approximation

$\frac{3.5625 + 3.625}{2} = 3.59375$

$\mathrm{f}(3.59375) = 3.59375^2 - 13 = -0.084960937$

(b) $x_{n+1} = x_n - \dfrac{f(x_n)}{f'(x_n)}$

Let $x_n = 3.59375$

$f(x) = x^2 - 13 \Rightarrow f(3.59375) = 3.59375^2 - 13 = -0.084960937$

$f'(x) = 2x \quad f'(3.59375) = 3 \times 3.59355 = 7.187$

$x_{n+1} = 3.59375 - \dfrac{-0.084960937}{7.187} = 3.605506796.$

---

9. (a) $\sum_{r=1}^{n} r(r+1)(r+2) = \sum_{r=1}^{n} (r^3 + 3r^2 + 2r)$

$$= \sum_{r=1}^{n} r^3 + 3\sum_{r=1}^{n} r^2 + 2\sum_{r=1}^{n} r$$

where $\quad r(r+1)(r+2) = r(r^2 + 3r + 2) = r^3 + 3r^2 + 2r$

$$\sum_{r=1}^{n} r(r+1)(r+2) = \frac{n^2(n+1)^2}{4} + \frac{n(n+1)(2n+1)}{2} + n(n+1)$$

$$= \frac{1}{4}\left[n^2(n+1)^2 + 2n(n+1)(2n+1) + 4n(n+1)\right]$$

$$= \frac{n(n+1)}{4}\left[n(n+1) + 2(2n+1) + 4\right] = \frac{n(n+1)}{4}(n^2 + 5n + 6)$$

$$= \frac{1}{4}n(n+1)(n+2)(n+3).$$

(b) $\sum_{r=11}^{20} r(r+1)(r+2) = \sum_{r=1}^{20} r(r+1)(r+2) - \sum_{r=1}^{10} r(r+1)(r+2)$

$$= \frac{1}{4}20 \times 21 \times 22 \times 23 - \frac{1}{4}10 \times 11 \times 12 \times 13$$

$$= 53130 - 4290 = 48840.$$

Examination Test Papers.

GCE Advanced Level.

Further Pure Mathematics FP1

Test Paper 4

Time: 1 hour and 30 minutes.

Instructions and Information

Candidates may use any calculator allowed by the regulations of their Examination Board.

Full marks are awarded for correct answers to ALL questions.

This paper has nine questions.

You can start working with any question and you must label clearly all parts.

1. Find the condition that the roots of the equation $ax^2 + bx + c = 0$ are reciprocals. **(5)**

---

2. The complex numbers z_1 and z_2 are given by

$$z_1 = 2 + 3i$$
$$z_2 = 4 + 5i.$$

(a) Find $\frac{z_1}{z_2}$ in the form $a + ib$, where a and b are real rational numbers **(3)**

(b) Find the arg $\left(\frac{z_2}{z_1}\right)$. **(3)**

---

3. Show that $\sum_{r=1}^{n} r = 1 + 2 + 3 + \cdots + n = \frac{n(n+1)}{2}$

(i) by considering it as an arithmetic series. **(3)**

(ii) by using the difference squares. **(5)**

---

4. Use the method of mathematical induction to show that $2^{3n} - 1$ is divisible by 7. **(8)**

---

5. The transformation matrix for rotating a positive vector $\begin{pmatrix} x \\ y \end{pmatrix}$ clock wise about the origin at an angle θ is $\begin{pmatrix} \cos\theta & \sin\theta \\ -\sin\theta & \cos\theta \end{pmatrix}$.

Determine the new position vectors when the position vector $\overrightarrow{OA} = \begin{pmatrix} 1 \\ 0 \end{pmatrix}$ is rotated about the origin through the following angles:

(i) 90° (ii) 180° (iii) 270° and (iv) 360°. **(8)**

---

6. (a) A square S is given by the vertices $A(-2, 0)$, $B(0, 2)$, $C(2, 0)$ and $D(0, -2)$.

Find the vertices of the image of S under the transformation $\begin{pmatrix} 1 & 1 \\ 1 & -1 \end{pmatrix}$.

Determine the area of the image. **(4)**

(b) The right angled triangle T, with vertices (4, 2), (6, 2) and (6, 4).

Find the vertices of the image of T under the transformation $\begin{pmatrix} 1 & 1 \\ \frac{1}{2} & 1 \end{pmatrix}$ and find its area. **(4)**

(c) Draw the graphs for the square S and the triangles. **(4)**

---

7. Find the coordinates of the point of intersection of the normals at the points P $(ap^2, 2ap)$ and Q $(aq^2, 2aq)$ of the parabola. $y^2 = 4ax$. **(12)**

---

8. Find the inverse matrix of $\mathbf{A}\begin{pmatrix} 3 & -1 \\ -4 & 3 \end{pmatrix}$,

explaining in details the intermediate steps. Hence show that

$\mathbf{A}^{-1}\mathbf{A} = \mathbf{A}\mathbf{A}^{-1} = \mathbf{I} = \begin{pmatrix} 1 & 0 \\ 0 & 1 \end{pmatrix}$. **(8)**

---

9. if α and β are the roots of the equation $3x^2 - 5x - 7 = 0$.

Find the values of the following:

(i) $\dfrac{1}{\alpha} + \dfrac{1}{\beta}$ (ii) $\dfrac{1}{\alpha^3} - \dfrac{1}{\beta^3}$. **(8)**

TOTAL FOR PAPER: 75 MARKS

Examination Test Papers.

GCE Advanced Level.

Test Paper 4 Solutions

Further Pure Mathematics FP1

1. If α and β are the roots of the quadratic equation $ax^2 + bx + c = 0$, then $\alpha = \frac{1}{\beta}$ or $\beta = \frac{1}{\alpha}$.

$$S = \text{sum} = \alpha + \beta = -\frac{b}{a} = \alpha + \frac{1}{\alpha} = \frac{\alpha^2 + 1}{\alpha} \qquad \ldots (1)$$

$$P = \text{product} = \alpha\beta = \frac{c}{a} = \alpha\frac{1}{\alpha} = 1 \Rightarrow c = a$$

$$\alpha^2 + 1 + \alpha\left(\frac{b}{a}\right) = 0 \qquad \text{from (1)}$$

$$\alpha^2 + \frac{b}{a}\alpha + 1 = 0$$

$$\alpha = \frac{1}{2}\left[-\frac{b}{a} \pm \sqrt{\left(\frac{b}{a}\right)^2 - 4 \times 1 \times 1}\right] \quad \Rightarrow \quad \boxed{\frac{b^2}{4a^2} \geq 1}$$

---

2. (a) $\frac{z_1}{z_2} = \frac{2+3i}{4+5i} \times \frac{4-5i}{4-5i} = \frac{8 + 12i - 10i + 15}{16 - 25i^2}$

$$= \frac{23}{41} + \frac{2}{41}i$$

$a = \frac{23}{41}$ and $b = \frac{2}{41}$ are real rational numbers.

(b) $\arg\left(\frac{z_2}{z_1}\right) = \arg z_2 - \arg z_1$

$$= \tan^{-1}\frac{5}{4} - \tan^{-1}\frac{3}{2} = 51.3° - 56.3° = -5°.$$

---

3. (i) $1 + 2 + 3 + \cdots + (n-2) + (n-1) + n = \sum_{r=1}^{n} r \qquad \ldots (1)$

$$n + (n-1) + (n-2) + \cdots + 3 + 2 + 1 = \sum_{r=1}^{n} r \qquad \ldots (2)$$

Adding (1) and (2)

$$(n+1)n = 2\sum_{r=1}^{n} r \quad \Rightarrow \quad \sum_{r=1}^{n} r = \frac{n(n+1)}{2}.$$

(ii)

$$
\begin{aligned}
n^2 - (n-1)^2 &= n^2 - n^2 + 2n - 1 && = 2n - 1 \\
(n-1)^2 - (n-2)^2 &= && = 2(n-1) \\
&\vdots \\
2^2 - 1^2 &= && = 2 \times 2 - 1 \\
1^2 - 0^2 &= && = 2 \times 1 - 1 \\
\hline
n^2 &= && = 2\sum_{r=1}^{n} r - n
\end{aligned}
$$

$$\therefore 2\sum_{r=1}^{n} r = n + n^2 = n(n+1)$$

$$\therefore \sum_{r=1}^{n} r = \frac{n(n+1)}{2}.$$

---

4. Let $f(n) = 2^{3n} - 1$ where $n \in \mathbb{Z}^+$,

$f(1) = 8 - 1 = 7$, which is divisible by 7.

$f(2) = 2^6 - 1 = 63$ which is also divisible by 7.

$f(k) = 2^{3k} - 1$ would also be divisible by 7

$f(k+1) = 2^{3(k+1)} - 1$

$f(k+1) - f(k) = 2^{3k+3} - 1 - (2^{3k} - 1) = 2^3.2^{3k} - 2^{3k} = 2^{3k}(2^3 - 1)$

$= 7 \times 2^{3k}$ which is divisible by 7.

$\therefore f(n)$ is divisible by 7 when $n = k + 1$.

---

5. (i) $\begin{pmatrix} \cos 90^\circ & \sin 90^\circ \\ -\sin 90^\circ & \cos 90^\circ \end{pmatrix}\begin{pmatrix} 1 \\ 0 \end{pmatrix} = \begin{pmatrix} 0 & 1 \\ -1 & 0 \end{pmatrix} = \begin{pmatrix} 0 \\ -1 \end{pmatrix}$

(ii) $\begin{pmatrix} \cos 180^\circ & \sin 180^\circ \\ -\sin 180^\circ & \cos 180^\circ \end{pmatrix}\begin{pmatrix} 1 \\ 0 \end{pmatrix} = \begin{pmatrix} -1 & 0 \\ 0 & -1 \end{pmatrix}\begin{pmatrix} 1 \\ 0 \end{pmatrix} = \begin{pmatrix} -1 \\ 0 \end{pmatrix}$

(iii) $\begin{pmatrix} \cos 270^\circ & \sin 270^\circ \\ -\sin 270^\circ & \cos 270^\circ \end{pmatrix}\begin{pmatrix} 1 \\ 0 \end{pmatrix} = \begin{pmatrix} 0 & -1 \\ 1 & 0 \end{pmatrix}\begin{pmatrix} 1 \\ 0 \end{pmatrix} = \begin{pmatrix} 0 \\ 1 \end{pmatrix}$

(iv) $\begin{pmatrix} \cos 360^\circ & \sin 360^\circ \\ -\sin 360^\circ & \cos 360^\circ \end{pmatrix}\begin{pmatrix} 1 \\ 0 \end{pmatrix} = \begin{pmatrix} 1 & 0 \\ 0 & 1 \end{pmatrix}\begin{pmatrix} 1 \\ 0 \end{pmatrix} = \begin{pmatrix} 1 \\ 0 \end{pmatrix}.$

---

6. (a) $\begin{pmatrix} 1 & 1 \\ 1 & -1 \end{pmatrix}\begin{pmatrix} -2 & 0 & 2 & 0 \\ 0 & 2 & 0 & -2 \end{pmatrix} = \begin{pmatrix} -2 & 2 & 2 & -2 \\ -2 & -2 & 2 & 2 \end{pmatrix}$.

$$\begin{matrix} \text{A} & \text{B} & \text{C} & \text{D} \end{matrix} \qquad \begin{matrix} \text{A}' & \text{B}' & \text{C}' & \text{D}' \end{matrix}$$

Area of image = area of object $\times |\det(\text{M})|$

$$|\det(\text{M})| = \left| \begin{vmatrix} 1 & 1 \\ 1 & -1 \end{vmatrix} \right| = |-1 - 1| = |-2| = 2$$

area of image $= 8 \times 2 = 16$ s.u.

area of object $= \dfrac{2 \times 2}{2} \times 4 = 8$ s.u.

(b) $\begin{pmatrix} 1 & 1 \\ \frac{1}{2} & -1 \end{pmatrix}\begin{pmatrix} 4 & 6 & 6 \\ 2 & 2 & 4 \end{pmatrix} = \begin{pmatrix} 6 & 8 & 10 \\ 4 & 5 & 7 \end{pmatrix}$

$$\begin{matrix} \text{E} & \text{F} & \text{G} \end{matrix} \qquad \begin{matrix} \text{E}' & \text{F}' & \text{G}' \end{matrix}$$

area of image = area of object $\times |\det(\text{M})|$

$$= \frac{2 \times 2}{2} \times \begin{vmatrix} 1 & 1 \\ \frac{1}{2} & 1 \end{vmatrix} = 2 \times \frac{1}{1} = 1 \text{ s.u.}$$

(c)

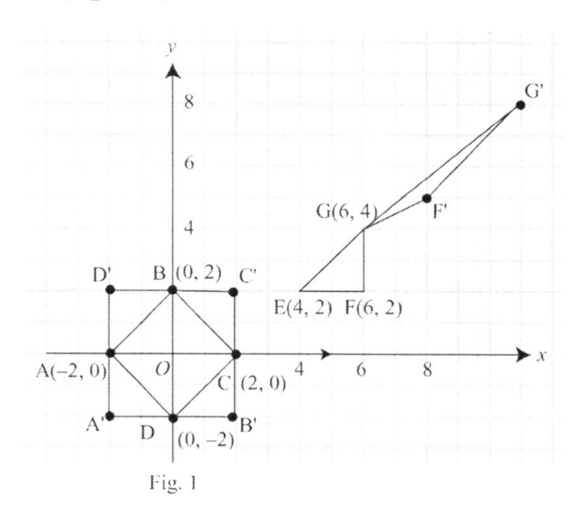

Fig. 1

---

7. $y^2 = 4ax$

differentiating with respect to x $\quad 2y\dfrac{dy}{dx} = 4a \quad \Rightarrow \quad \dfrac{dy}{dx} = \dfrac{2a}{y} = \dfrac{2a}{2ap} = \dfrac{1}{p}$

the gradient of the tangent, and the gradient of the normal is $-p$ and therefore the gradient of the tangent and normal at Q are $\dfrac{1}{q}$ and $-q$ respectively.

If $y = mx + c$, the equations of the normals at P and Q are:

$y = -px + c \quad \Rightarrow \quad 2ap = -pap^2 + c \quad \Rightarrow \quad c = 2ap + ap^3,$

$\therefore \boxed{y = -px + 2ap + ap^3}$

and that at Q is $y = -qx + 2aq + aq^3$

$$y + px = 2ap + ap^3 \qquad \ldots(1)$$

$$y + qx = 2aq + aq^3 \qquad \ldots(2)$$

$(1) - (2) \quad x(p - q) = 2ap + ap^3 - 2aq - aq^3$

$$x = \frac{2a(p - q) + a(p^3 - q^3)}{p - q}$$

$$\boxed{x = 2a + a(p^2 + pq + q^2)}$$

$y + px = 2ap + ap^3$

$$y = -2ap - ap(p^2 + pq + q^2) + 2ap + ap^3$$

$$y = -apq(p + q)$$

$\therefore \left[2a + a(p^2 + pq + q^2, -apq(p + q)\right].$

---

8. Step I

The minors of $\mathbf{A} = \begin{pmatrix} 3 & -4 \\ -1 & 3 \end{pmatrix}$.

The cofactors of $\mathbf{A} = \mathbf{A}^* = \begin{pmatrix} 3 & 4 \\ 1 & 3 \end{pmatrix}$.

Step II

$\mathbf{A}^{*\mathrm{T}} = \begin{pmatrix} 3 & 1 \\ 4 & 3 \end{pmatrix}$ = the adjoining matrix of $\mathbf{A}$

Step III

$\mathbf{A}^{-1} = \dfrac{\mathbf{A}^{*\mathrm{T}}}{|\mathbf{A}|} \qquad |\mathbf{A}| = \begin{vmatrix} 3 & -1 \\ -4 & 3 \end{vmatrix} \begin{matrix} = 3 \times 3 - (-1)(-4) \\ = 9 - 4 = 5 \end{matrix}$

$$\mathbf{A}^{-1} = \frac{1}{5}\begin{pmatrix} 3 & 1 \\ 4 & 3 \end{pmatrix} = \begin{pmatrix} \frac{3}{5} & \frac{1}{5} \\ \frac{4}{5} & \frac{3}{5} \end{pmatrix}$$

$$\mathbf{A}^{-1}\mathbf{A} = \begin{pmatrix} \frac{3}{5} & \frac{1}{5} \\ \frac{4}{5} & \frac{3}{5} \end{pmatrix} \begin{pmatrix} 3 & -1 \\ -4 & 3 \end{pmatrix} = \begin{pmatrix} \frac{3}{5} + 3 + \frac{1}{5} \times -4 & \frac{3}{5} \times -1 + \frac{1}{5} \times 3 \\ \frac{4}{5} \times 3 + \frac{3}{5} \times -4 & \frac{4}{5} \times -1 + \frac{3}{5} \times 3 \end{pmatrix}$$

$$= \begin{pmatrix} 1 & 0 \\ 0 & 1 \end{pmatrix} = \mathbf{I} \quad \text{unit matrix}$$

$$\mathbf{A}\mathbf{A}^{-1} = \begin{pmatrix} 3 & -1 \\ -4 & 3 \end{pmatrix} = \begin{pmatrix} \frac{3}{5} & \frac{1}{5} \\ \frac{4}{5} & \frac{3}{5} \end{pmatrix} = \begin{pmatrix} 1 & 0 \\ 0 & 1 \end{pmatrix} = \mathbf{I} \quad \text{unit matrix.}$$

---

9. (i) $3x^2 - 5x + 7 = 0$

$$\alpha + \beta = \frac{5}{3} \qquad \alpha\beta = \frac{7}{3}$$

$$\frac{1}{\alpha} + \frac{1}{\beta} = \frac{\alpha + \beta}{\alpha\beta} = \frac{\frac{5}{3}}{\frac{7}{3}} = \frac{5}{7} \quad \text{real}$$

(ii) $$\frac{1}{\alpha^3} - \frac{1}{\beta^3} = \frac{\beta^3 - \alpha^3}{\alpha^3\beta^3} = \frac{(\beta - \alpha)(\beta^2 + \alpha\beta + \alpha^2)}{\alpha^3\beta^3}$$

$$= \frac{\beta - \alpha}{\alpha^3\beta^3}\left[(\alpha + \beta)^2 - 2\alpha\beta + \alpha\beta\right]$$

$$= \frac{\beta - \alpha}{\left(\frac{7}{3}\right)^3}\left[\left(\frac{5}{3}\right)^2 - \frac{7}{3}\right] = \frac{(\beta - \alpha)\frac{4}{9}}{\frac{343}{27}} = (\beta - \alpha)\frac{12}{343}$$

but $(\beta - \alpha)^2 = \beta^2 - 2\alpha\beta + \alpha^2 = (\beta + \alpha)^2 - 4\alpha\beta$

$$\beta - \alpha = \pm\sqrt{(\beta + \alpha)^2 - 4\alpha\beta} = \pm\sqrt{\frac{25}{9} - 4 \times \frac{7}{3}} = \pm\sqrt{\frac{-59}{9}} = \pm i\frac{\sqrt{59}}{3}$$

$$\therefore \frac{1}{\alpha^3} - \frac{1}{\beta^3} = \frac{\beta^3 - \alpha^3}{\alpha^3\beta^3} = \frac{(\beta - \alpha)(\beta^2 + \alpha\beta + \alpha^2)}{(\alpha\beta)^3} = \pm\frac{\sqrt{59}}{3}\frac{i\frac{4}{9}}{\left(\frac{7}{3}\right)^3}$$

$$= \pm\frac{\sqrt{59}}{3} \times \frac{4}{9} \times \frac{27}{343}i = \pm 0.09i \quad \text{complex.}$$

Examination Test Papers.

GCE Advanced Level.

Further Pure Mathematics FP1

Test Paper 5

Time: 1 hour and 30 minutes.

Instructions and Information

Candidates may use any calculator allowed by the regulations of their Examination Board.

Full marks are awarded for correct answers to ALL questions.

This paper has nine questions.

You can start working with any question and you must label clearly all parts.

1. If α and β are the roots of the eqaution $-7x^2 + 9x - 7 = 0$, find the values

(i) $\dfrac{\alpha}{\beta} + \dfrac{\beta}{\alpha}$ (ii) $\dfrac{1}{\alpha^3} + \dfrac{1}{\beta^3}$. **(6)**

---

2. Find expressions for the following summations:

(i) $\sum_{r=1}^{n} r(r+1)$ (ii) $\sum_{r=1}^{n} r(r-1)$ (iii) $\sum_{r=1}^{n} r(r+3)$. **(9)**

---

3. Solve the simultaneous equations

$$3x - y = 9$$
$$-4x + 3y = -7$$

by means of matrices. **(8)**

---

4. If $\mathbf{A} = \begin{pmatrix} 6 & 7 \\ 8 & 9 \end{pmatrix}$, $\mathbf{B} = \begin{pmatrix} 10 & 11 \\ 12 & 13 \end{pmatrix}$ and $\mathbf{C} = \begin{pmatrix} 0 & 1 \\ -1 & 0 \end{pmatrix}$.

Find

(i) **BC** (ii) **CB** (iii) **ABC** **(6)**

show that $\mathbf{AB} \neq \mathbf{BA}$. **(2)**

---

5. $f(x) = x^3 - 4x^2 + 13x + 50$.

Given that $x = -2$ is a solution of $f(x) = 0$, solve this equation completely. **(6)**

---

6. Find the modulus and argument of the complex number $z = \dfrac{3 - 4i}{5 + 12i}$. **(8)**

---

7. $f(x) = 3^x + 2x + 4$.

The equation $f(x) = 0$ has a root x in the interval $[-3, -2]$.

Use interval bisection for the end points to

show that after 4 approximations $x = -2.03125$. **(10)**

---

8. Use the method of mathematical induction to prove that $27^n + 3$,

for all positive integers, is divisible by 3. **(8)**

---

9. Use the method of mathematical induction to show that

$$\sum_{r=1}^{n} r^4 = \frac{1}{30}n(n+1)(6n^3 + 9n^2 + n - 1).$$ **(12)**

TOTAL FOR PAPER: 75 MARKS

Examination Test Papers.

GCE Advanced Level.

Test Paper 5 Solutions

Further Pure Mathematics FP1

1. (i) $-7x^2 + 9x - 7 = 0$

$$\alpha + \beta = -\frac{9}{-7} = \frac{9}{7} \qquad \alpha\beta = \frac{-7}{-7} = 1$$

$$\frac{\alpha}{\beta} + \frac{\beta}{\alpha} = \frac{\alpha^2 + \beta^2}{\alpha\beta} = \frac{(\alpha + \beta)^2 - 2\alpha\beta}{\alpha\beta} = \frac{(\alpha + \beta)^2}{\alpha\beta} - 2$$

$$= \frac{\left(\frac{9}{7}\right)^2}{1} - 2 = \frac{81}{49} - 2 = \frac{81 - 98}{49} = -\frac{17}{49}.$$

(ii) $$\frac{1}{\alpha^3} + \frac{1}{\beta^3} = \frac{\alpha^3 + \beta^3}{\alpha^3\beta^3} = \frac{(\alpha + \beta)^3 - 3\alpha\beta(\alpha + \beta)}{\alpha^3\beta^3}$$

$$= \frac{\left(\frac{9}{7}\right)^3 - 3 \times 1 \times \frac{9}{7}}{1} = \frac{729}{343} - \frac{27 \times 49}{7 \times 49}$$

$$= \frac{729}{343} - \frac{1323}{343} = -\frac{594}{343}.$$

---

2. (i) $$\sum_{r=1}^{n} r(r + 1) = \sum_{r=1}^{n} r^2 + \sum_{r=1}^{n} r$$

$$= \frac{n(n + 1)(2n + 1)}{6} + \frac{n(n + 1) \times 3}{2 \times 3}$$

$$= \frac{n(n + 1)(2n + 4)}{6} = \frac{n(n + 1)(n + 2)}{3}$$

(ii) $$\sum_{r=1}^{n} r(r - 1) = \sum_{r=1}^{n} r^2 - \sum_{r=1}^{n} r = \frac{n(n + 1)(2n + 1)}{6} - \frac{n(n + 1)}{2 \times 3} \times 3$$

$$= \frac{n(n + 1)2(n - 1)}{6} = \frac{n(n + 1)(n - 1)}{3}$$

(iii) $$\sum_{r=1}^{n} r(r + 3) = \sum_{r=1}^{n} r^2 + 3\sum_{r=1}^{n} r = \frac{n(n + 1)(2n + 1)}{6} + \frac{3n(n + 1)}{2 \times 3} \times 3$$

$$= \frac{n(n + 1)(n + 5)}{3}.$$

3.

$$3x - y = 9$$
$$-4x + 3y = -7$$

$$\begin{pmatrix} 3 & -1 \\ -4 & 3 \end{pmatrix}\begin{pmatrix} x \\ y \end{pmatrix} = \begin{pmatrix} 9 \\ -7 \end{pmatrix} \qquad \ldots(1)$$

Let $\mathbf{A} = \begin{pmatrix} 3 & -1 \\ -4 & 3 \end{pmatrix}$ the minors of $\mathbf{A} = \begin{pmatrix} 3 & -4 \\ -1 & 3 \end{pmatrix}$

$\mathbf{A}^* = \begin{pmatrix} 3 & 4 \\ 1 & 3 \end{pmatrix}$ = the cofactors of $\mathbf{A}$

$\mathbf{A}^{*\mathrm{T}} = \begin{pmatrix} 3 & 1 \\ 4 & 3 \end{pmatrix}$ and hence $\mathbf{A}^{-1} = \dfrac{\mathbf{A}^{*\mathrm{T}}}{|\mathbf{A}|} = \begin{pmatrix} \frac{3}{5} & \frac{1}{5} \\ \frac{4}{5} & \frac{3}{5} \end{pmatrix}$

where $|\mathbf{A}| = 3 \times 3 - (-1)(-4) = 9 - 4 = 5.$

Pre-multiplying each side of (1) by $\mathbf{A}^{-1}$

$$\begin{pmatrix} \frac{3}{5} & \frac{1}{5} \\ \frac{4}{5} & \frac{3}{5} \end{pmatrix}\begin{pmatrix} 3 & -1 \\ -4 & 3 \end{pmatrix}\begin{pmatrix} x \\ y \end{pmatrix} = \begin{pmatrix} \frac{3}{5} & \frac{1}{5} \\ \frac{4}{5} & \frac{3}{5} \end{pmatrix}\begin{pmatrix} 9 \\ -7 \end{pmatrix}$$

$$\mathbf{I}\begin{pmatrix} x \\ y \end{pmatrix} = \begin{pmatrix} \frac{27}{5} - \frac{7}{5} \\ \frac{36}{5} - \frac{21}{5} \end{pmatrix} = \begin{pmatrix} 4 \\ 3 \end{pmatrix}$$

$\therefore \boxed{x = 4} \quad \boxed{y = 3}$

---

4. (i) $\mathbf{BC} = \begin{pmatrix} 10 & 11 \\ 12 & 13 \end{pmatrix}\begin{pmatrix} 0 & 1 \\ -1 & 0 \end{pmatrix} = \begin{pmatrix} -11 & 10 \\ -13 & 12 \end{pmatrix}$

(ii) $\mathbf{CB} = \begin{pmatrix} 0 & 1 \\ 1 & 0 \end{pmatrix}\begin{pmatrix} 10 & 11 \\ 12 & 13 \end{pmatrix} = \begin{pmatrix} 12 & 13 \\ -10 & -11 \end{pmatrix}$

(iii) $\mathbf{ABC} = \begin{pmatrix} 6 & 7 \\ 8 & 9 \end{pmatrix} \begin{pmatrix} 10 & 11 \\ 12 & 13 \end{pmatrix} \begin{pmatrix} 0 & 1 \\ -1 & 0 \end{pmatrix} = \begin{pmatrix} 144 & 157 \\ 188 & 205 \end{pmatrix} \begin{pmatrix} 0 & 1 \\ -1 & 0 \end{pmatrix}$

$$= \begin{pmatrix} -157 & 144 \\ -205 & 188 \end{pmatrix}$$

$$\mathbf{AB} = \begin{pmatrix} 144 & 157 \\ 188 & 205 \end{pmatrix} \qquad \mathbf{BA} \begin{pmatrix} 10 & 11 \\ 12 & 13 \end{pmatrix} \begin{pmatrix} 6 & 7 \\ 8 & 9 \end{pmatrix} = \begin{pmatrix} 144 & 169 \\ 176 & 201 \end{pmatrix}$$

$\therefore \mathbf{AB} \neq \mathbf{BA}$ not commutative.

---

5. $f(-2) = -8 - 16 - 26 + 50 = 0$

therefore $x + 2$ is a factor of $f(x)$.

$$\begin{array}{r|l} & x^2 - 6x + 25 \\ \hline x+2 & x^3 - 4x^2 + 13x + 50 \\ & \underline{x^3 + 2x^2} \\ & \quad - 6x^2 + 13x + 50 \\ & \quad \underline{- 6x^2 - 12x} \\ & \qquad 25x + 50 \\ & \qquad \underline{25x + 50} \\ & \qquad\qquad 0 \end{array}$$

$f(x) = (x + 2)(x^2 - 6x + 25) = 0$

$x^2 - 6x + 25 = 0 \quad \Rightarrow \quad x = \dfrac{6 \pm \sqrt{36 - 100}}{2} = \dfrac{6 \pm i8}{2}$

$\therefore x = -2, \quad x = 3 + i4, \quad x = 3 - 4i.$

---

6. $z = \dfrac{3 - 4i}{5 + 12i} = \dfrac{z_1}{z_2} \qquad |z| = \dfrac{|z_1|}{|z_2|} = \dfrac{\sqrt{(3)^2 + (-4)^2}}{\sqrt{5^2 + 12^2}} = \dfrac{5}{13}.$

Alternatively

$$z = \frac{3 - 4i}{5 + 12i} \times \frac{5 - 12i}{5 - 12i} = \frac{15 - 20i - 36i - 48}{5^2 - 12^2 i^2} = \frac{-33 - 56i}{169}$$

$$|z| = \sqrt{\left(\frac{-33}{169}\right)^2 + \left(\frac{-56}{169}\right)^2} = \frac{1}{169}\sqrt{1089 + 3136} = \frac{65}{169} = \frac{5}{13}.$$

The argument

$$\arg z = 180° + \tan^{-1} \frac{56}{33}$$

$$= 180° + 59.48976259°$$

$$= 180° + 59°29' = 239°29'.$$

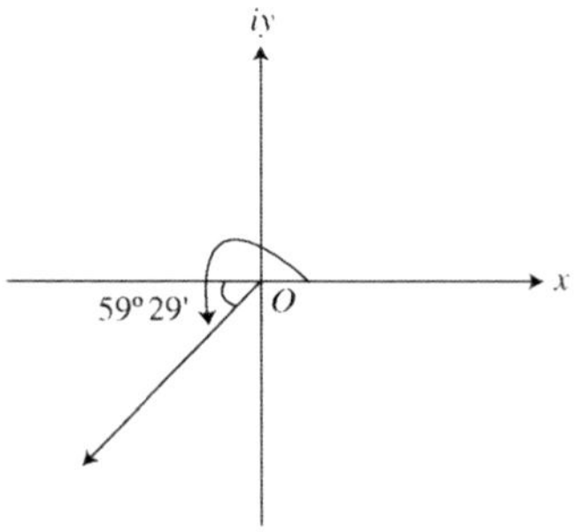

Alternatively

$$\arg z = \arg (3 - 4i) - \arg (5 + 12i) = -53.1° - 67.4°$$

$$= -120.5° = 239.5°$$

---

7. $$f(x) = 3^x + 2x + 4$$

$$f(-3) = 3^{-3} + 2(-3) + 4 = \frac{1}{27} - 2 = -\frac{53}{27}$$

$$f(-2) = 3^{-2} + 2(-2) + 4 = \frac{1}{9}$$

$$\frac{-2-3}{2} = -2.5 \quad \text{the mid-point}$$

$$f(-2.5) = 3^{-2.5} + 2(-2.5) + 4 = \frac{1}{3^{2.5}} - 1 = -0.93584997$$

first approximation

$$\frac{-2.5-2}{2} = -2.25 \quad \text{the mid-point}$$

$$f(-2.125) = 3^{-2.25} + 2(-2.25) + 4 = -0.415573812$$

second approximation

$$\frac{-2.25-2}{2} = -2.125$$

$$f(-2.125) = 3^{2.125} + 2(-2.125) + 4 = -0.15314605$$

third approximation

$$\frac{-2.125-2}{2} = -2.0625$$

$$f(-2.0625) = -0.146262109$$

fourth approximation

$$f(-2.03125) = 0.044861223$$

$$\therefore x = -2.03125.$$

---

8. $f(x) = 27^n + 3$

$f(1) = 27 + 3$ it is divisible by 3

$f(2) = 27^2 + 3 = 729 + 3 = 732$ it is also divisible by 3

$f(k) = 27^k + 3$ it would be divisible by 3

$f(k+1) = 27^{k+1} + 3 = 27 \times 27^{k+3} = 3(9 \times 27^k + 1)$

which is divisible by 3 and therefore $f(n)$ is divisible by 3.

---

9. $\sum_{r=1}^{n} r^4 = \frac{n}{30}(n+1)(6n^3 + 9n^2 + n - 1).$

For $n = 1$ $\quad 1^4 = \frac{1}{30}(1+1)(6+9+1-1) = 1$ LHS = RHS is true

For $n = 2$ $\quad 1^4 + 2^4 = \frac{2}{30}(2+1)(48+36+2-1) = \frac{2 \times 3 \times 85}{30} = 17$

it would be true for $n = k$

$$1^4 + 2^4 + 3^4 + \cdots + k^4 = \frac{k}{30}(k+1)(6k^3 + 9k^2 + k - 1) \qquad \ldots(1)$$

Adding the $(k+1)$th term to both sides of (1)

$$1^4 + 2^4 + \cdots + k^4 + (k+1)^4 = \frac{k(k+1)}{30}(6k^3 + 9k^2 + k - 1) + (k+1)^4$$

$$= (k+1)\left[\frac{k}{30}(6k^3 + 9k^2 + k - 1) + (k+1)^3\right]$$

$$= (k+1)\left[\frac{k}{30}(6k^3 + 9k^2 + k - 1) + k^3 + 3k^2 + 3k + 1\right]$$

$$= \frac{(k+1)}{30}(6k^4 + 9k^3 + k^2 - k + 30k^3 + 90k^2 + 90k + 30$$

$$= \frac{(k+1)}{30}(6k^4 + 39k^3 + 91k^2 + 89k + 30) \qquad \ldots(2)$$

Let $f(k) = 6k^4 + 39k^3 + 91k^2 + 89k + 30$

$f(-2) = 96 - 312 + 364 - 178 + 30 = 0 \qquad \therefore x + 2$ is a factor

$f(k) = (k+2)(6k^3 + 27k^2 + 37k + 15).$

The right hand side of (2) is $(k+1)(k+2)(6k^3 + 27k^2 + 37k + 15).$

$\therefore$ it is true for n.

Examination Test Papers.

GCE Advanced Level.

Further Pure Mathematics FP1

Test Paper 6

Time: 1 hour and 30 minutes.

Instructions and Information

Candidates may use any calculator allowed by the regulations of their Examination Board.

Full marks are awarded for correct answers to ALL questions.

This paper has nine questions.

You can start working with any question and you must label clearly all parts.

1. Use the method of mathematical induction to show that

$$\sum_{r=1}^{n} r(2^r) = 2^{n+1}(n-1) + 2.$$ **(6)**

---

2. A right angled triangle T has vertices A (1, 0), B (2, 0) and C (2, 2)

when T is transformed by the matrix $\mathbf{P} = \begin{pmatrix} 0 & 1 \\ 1 & 0 \end{pmatrix}$, the image is T′.

(a) Find the coordinates of the vertices of T′.

(b) Describe fully the transformation represented by P.

Sketch these triangles.

The matrices $\mathbf{R} = \begin{pmatrix} 1 & 2 \\ 3 & 4 \end{pmatrix}$ and $\mathbf{S} = \begin{pmatrix} 5 & -6 \\ -7 & 8 \end{pmatrix}$ represent two transformations.

When T′ is transformed by the image **RS**, the image is T".

Find **RS**.

(d) Find the determinant of RS.

(e) Using your answer to part (d), find the area of T". **(10)**

---

3. Rotate clockwise the right angled triangle A(2, 1), B (3, 1) and C (3, 3) by

(a) $\theta = 30°$ (b) $45°$ (c) $90°$ (d) $180°$. **(8)**

---

4. The roots of a quadratic equation $x^2 + 2x + 3 = 0$ are α and β.

Find the quadratic equation whose roots are $\alpha + \dfrac{2}{\beta}$ and $\beta + \dfrac{2}{\alpha}$. **(8)**

---

5. Prove $z_1{}^* + z_2{}^* = (z_1 + z_2)^*$ and $(z_1 z_2)^* = z_1{}^* \, z_2{}^*$

where $z_1 = a_1 + b_1 i$ and $z_2 = a_2 + b_2 i$ where $(a_1 b_1, a_2, b_2 \in \mathbb{R})$. **(8)**

---

6. Sketch the curve $y = \dfrac{x}{1+x}$ by considering the gradient and stating the asymptotes. **(6)**

---

7. Prove that $\sum_{r=1}^{n} r^2 = \frac{n(n+1)(2n+1)}{6}$ using the difference of squares. **(12)**

---

8. Sketch the rectangular hyperbola $xy = c^2$.

Find the equations of the tangent and normal at P $(ct, \frac{c}{t})$.
Hence determine the coordinates of the points of the intersection with the axes. **(9)**

---

9. Solve the simultaneous equations

$$5x + 4y = -2$$
$$2x - 3y = 13$$

by means of matrices. **(8)**

TOTAL FOR PAPER: 75 MARKS

Examination Test Papers.

GCE Advanced Level.

Test Paper 6 Solutions

Further Pure Mathematics FP1

1. $\sum_{r=1}^{n} r\,2^r = 2^{n+1}(n-1) + 2.$

For $n = 1$, $\quad 1 \times 2^1 = 2^{1+1}(1-1) + 2 = 2.$

LHS = RHS $\quad$ it is true.

For $n = 2$, $\quad 1 \times 2 + 2 \times 2^2 = 2^{2+1}(2-1) + 2 = 10$

LHS = RHS $\quad$ it is also true.

For $n = k \quad 1 \times 2^1 + 2 \times 2^2 + 3 \times 2^3 + \ldots + k \times 2^k = 2^{k+1}(k-1) + 2 \quad \ldots(1)$

it would be true.

Adding the $(k+1)$th term to both sides of (1)

$$1 \times 2^1 + 2 \times 2^2 + 3 \times 2^3 + \ldots + k \times 2^k + (k+1) \times 2^{k+1} = 2^{k+1}(k-1) + 2 + (k+1) \times 2^{k+1}$$
$$= 2^{k+2}k + 2$$

$\therefore$ it is true for n.

---

2. a) $T' = \begin{pmatrix} 0 & 1 \\ 1 & 0 \end{pmatrix} \begin{pmatrix} 1 & 2 & 2 \\ 0 & 0 & 2 \end{pmatrix} = \begin{pmatrix} 0 & 0 & 2 \\ 1 & 2 & 2 \end{pmatrix}.$
$\qquad\qquad A' \; B' \; C'$

b) Reflection through $y = x$.

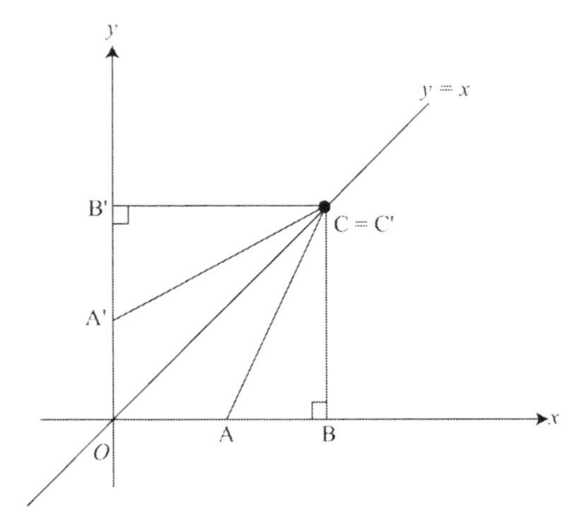

c) $\mathbf{RS} = \begin{pmatrix} 1 & 2 \\ 3 & 4 \end{pmatrix} \begin{pmatrix} 5 & -6 \\ -7 & 8 \end{pmatrix} = \begin{pmatrix} -9 & 10 \\ -13 & 14 \end{pmatrix}.$

d) Determinant of **RS**

$$\begin{vmatrix} -9 & 10 \\ -13 & 14 \end{vmatrix} = -9 \times 14 - 10 \times -13 = 4$$

e) Area T$''$ $= 4 \times$ area of T $= 4 \times 1 = 4$ s.u.

---

3. $\begin{pmatrix} \cos\theta & \sin\theta \\ -\sin\theta & \cos\theta \end{pmatrix} \begin{pmatrix} 2 & 3 & 3 \\ 1 & 1 & 3 \end{pmatrix}$

(a) $\theta = 30^\circ$

$$\begin{pmatrix} \frac{\sqrt{3}}{2} & \frac{1}{2} \\ -\frac{1}{2} & \frac{\sqrt{3}}{2} \end{pmatrix} \begin{pmatrix} 2 & 3 & 3 \\ 1 & 1 & 3 \end{pmatrix} = \begin{pmatrix} \sqrt{3} + \frac{1}{2} & \frac{3\sqrt{3}+1}{2} & \frac{3\sqrt{3}+3}{2} \\ -1 + \frac{\sqrt{3}}{2} & \frac{-3+\sqrt{3}}{2} & \frac{-3+3\sqrt{3}}{2} \end{pmatrix}$$

(b) $\theta = 45^\circ$

$$\begin{pmatrix} \frac{1}{\sqrt{2}} & \frac{1}{\sqrt{2}} \\ -\frac{1}{\sqrt{2}} & \frac{1}{\sqrt{2}} \end{pmatrix} \begin{pmatrix} 2 & 3 & 3 \\ 1 & 1 & 3 \end{pmatrix} = \begin{pmatrix} \frac{3}{\sqrt{2}} & \frac{4}{\sqrt{2}} & \frac{6}{\sqrt{2}} \\ -\frac{1}{\sqrt{2}} & \frac{-2}{\sqrt{2}} & 0 \end{pmatrix}$$

(c) $\theta = 90^\circ \begin{pmatrix} 0 & 1 \\ -1 & 0 \end{pmatrix} \begin{pmatrix} 2 & 3 & 3 \\ 1 & 1 & 3 \end{pmatrix} = \begin{pmatrix} 1 & 1 & 3 \\ -2 & -3 & -3 \end{pmatrix}$

(d) $\theta = 180^\circ$

$$\begin{pmatrix} -1 & 0 \\ 0 & -1 \end{pmatrix} \begin{pmatrix} 2 & 3 & 3 \\ 1 & 1 & 3 \end{pmatrix} = \begin{pmatrix} -2 & -3 & -3 \\ -1 & -1 & -3 \end{pmatrix}.$$

---

4. $x^2 + 2x + 3 = 0$

$\alpha + \beta = -2$ and $\alpha\beta = 3$.

The required quadratic equation is

$$x^2 - \left(\alpha + \frac{2}{\beta} + \beta + \frac{2}{\alpha}\right) x + \left(\alpha + \frac{2}{\beta}\right)\left(\beta + \frac{2}{\alpha}\right) = 0$$

$$x^2 - \left(\alpha + \beta + \frac{2\alpha + 2\beta}{\alpha\beta}\right) x + \alpha\beta + 2 + 2 + \frac{4}{\alpha\beta} = 0$$

$$x^2 - \left[-2 + \frac{2(-2)}{3}\right] x + 3 + 4 + \frac{4}{3} = 0$$

$$x^2 + \frac{10}{3}x + 7 + \frac{4}{3} = 0 \Rightarrow 3x^2 + 10x + 25 = 0.$$

---

5. $(z_1 + z_2)^* = (a_1 + b_1 i + a_2 + b_2 i)^* = [(a_1 + a_2) + (b_1 + b_2) i]^*$

$$= (a_1 + a_2) - (b_1 + b_2) i = a_1 - b_1 i + a_2 - b_2 i$$

$$= z_1{}^* + z_2{}^*$$

Prove $(z_1 z_2)^* = z_1{}^* z_2{}^*$

$$[(a_1 + b_1 i)(a_2 + b_2 i)]^* = (a_1 a_2 + b_1 a_2 i + a_1 b_2 i - b_1 b_2)^*$$

$$= [(a_1 a_2 - b_1 b_2) + (b_1 a_2 + a_1 b_2) i]^*$$

$$= a_1 a_2 - b_1 b_2 - (b_1 a_2 + a_1 b_2) i$$

$$z_1{}^* z_2{}^* = (a_1 - b_1 i)(a_2 - b_2 i)$$

$$= a_1 a_2 - a_2 b_1 i - a_1 b_2 i - b_1 b_2$$

$$= (a_1 a_2 - b_1 b_2) - (b_1 a_2 + a_1 b_2) i$$

$$\therefore (z_1 z_2)^* = z_1{}^* z_2{}^*.$$

---

6. $y = \dfrac{x}{1+x} \qquad \dfrac{dy}{dx} = \dfrac{1.(1+x) - x(1)}{(1+x)^2} = \dfrac{1}{(1+x)^2}$

the gradient is positive at any point of x.

To find the asymptotes.

$y = \dfrac{x}{x\left(1 + \frac{1}{x}\right)} = \dfrac{1}{1 + \frac{1}{x}}$ as $x \to \infty \quad y \to 1$

when $x = 0$, $\quad y = 0$

when $x \to -1 \quad y \to -\infty$

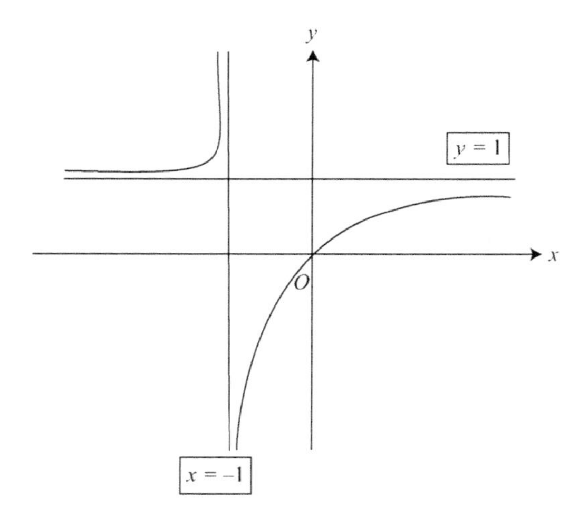

---

7. $n^3 - (n-1)^3 = n^3 - n^3 + 3n^2 - 3n + 1 \qquad = 3n^2 - 3n + 1$

$(n-1)^3 - (n-2)^3 = n^3 - 3n^2 + 3n - 1 - n^3 + 6n^2 - 12n + 8 = 3(n-1)^2 - 3(n-1) + 1$

$(n-2)^3 - (n-3)^3 = \qquad = 3(n-2)^2 - 3(n-2) + 1$

$\vdots$

$= 3 \times 1^2 - 3 \times 1 + 1$

$$\frac{1^3 - 0^3}{n^3} = \qquad = 3\sum_{r=1}^{n} r^2 - 3\sum_{r=1}^{n} r + n \qquad \ldots \textbf{(1)}$$

where $(a+b)^3 = a^3 + 3a^2b + 3ab^2 + b^3$

$(n-1)^3 = n^3 - 3n^2 + 3n - 1$

$(n-2)^3 = n^3 - 3 \times 2n^2 + 3(-2)^2 n - 8 = n^3 - 6n^2 + 12n - 8$

$$\begin{aligned}(n-1)^3 - (n-2)^3 &= n^3 - 3n^2 + 3n - 1 - n^3 + 6n^2 - 12n + 8\\ &= 3n^2 - 9n + 7 = 3(n-1)^2 - 3(n-1) + 1\\ &\qquad\qquad\qquad\;\; = 3n^2 - 6n + 3 - 3n + 3 + 1\\ &\qquad\qquad\qquad\;\; = 3n^2 - 9n + 7\end{aligned}$$

from (1)

$$n^3 = 3\sum_{r=1}^{n} r^2 - 3\sum_{r=1}^{n} r + n$$

$$3\sum_{r=1}^{n} r^2 = n^3 + \frac{3(n+1)n}{2} - n = \frac{2n^3 + 3n(n+1) - 2n}{2}$$

$$\sum_{r=1}^{n} r^2 = \frac{n}{6}\left(2n^2 + 3n + 1\right) = \frac{n}{6}\left(n+1\right)(2n+1)$$

$$\therefore \boxed{\sum_{r=1}^{n} r^2 = \frac{n(n+1)(2n+1)}{6}}.$$

---

8. $xy = c^2$ differentiating with respect to x

$$1y + x\frac{\mathrm{d}y}{\mathrm{d}x} = 0 \Rightarrow \frac{\mathrm{d}y}{\mathrm{d}x} = \frac{-y}{x} = \frac{-\frac{c}{t}}{ct} = -\frac{1}{t^2}$$

$$y = mx + k \Rightarrow y = -\frac{1}{t^2}x + k \Rightarrow \frac{c}{t} = -\frac{1}{t^2}x + \frac{2c}{t}$$

$$\text{since } k = \frac{c}{t} + \frac{c}{t} = \frac{2c}{t}$$

$\boxed{y = -\frac{1}{t^2}x + \frac{2c}{t}}$ ... (1) the equation of the tangent.

The gradient of the normal is t^2

$$y = t^2x + k \Rightarrow \frac{c}{t} = t^2ct + k \Rightarrow k = \frac{c}{t} - ct^3$$

$\boxed{y = t^2x + \frac{c}{t} - ct^3}$ ... (2) the equation of the normal

substituting the values in (1)

when $x = 0$, $y = \frac{2c}{t}$ and when $y = 0$, $x = 2ct$

$\boxed{A\left(0, \frac{2c}{t}\right)}$ $\boxed{B(2ct, 0)}$

substituting the values in (2)

when $x = 0$, $y = \frac{c}{t} - ct^3$, when $y = 0$, $t^2x = ct^3 - \frac{c}{t} \Rightarrow x = ct - \frac{c}{t^3}$

$\therefore \boxed{C\left(0, \frac{c}{t} - ct^3\right)}$ and $\boxed{D = \left(ct - \frac{c}{t^3}, 0\right)}$

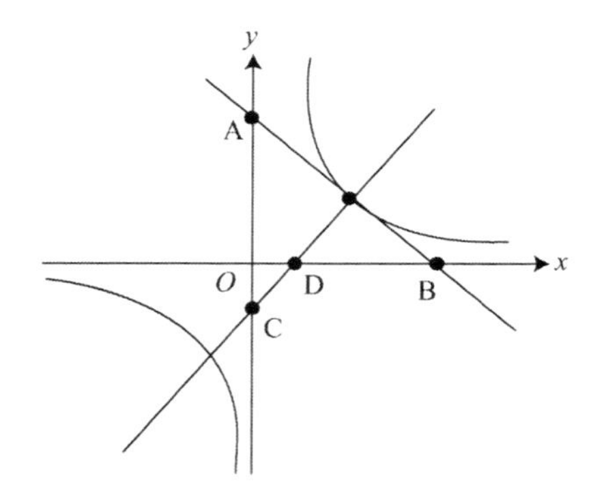

---

9. The equations can be written in matrix form as:

$$\begin{pmatrix} 5 & 4 \\ 2 & -3 \end{pmatrix} \begin{pmatrix} x \\ y \end{pmatrix} = \begin{pmatrix} -2 \\ 13 \end{pmatrix} \qquad \ldots (1)$$

Determine the inverse matrix of $\mathbf{A} = \begin{pmatrix} 5 & 4 \\ 2 & -3 \end{pmatrix}$

$$\text{the minors of } \mathbf{A} = \begin{pmatrix} -3 & 2 \\ 4 & 5 \end{pmatrix}$$

$$\text{the cofactors of } \mathbf{A} = \mathbf{A}^* = \begin{pmatrix} -3 & -2 \\ -4 & 5 \end{pmatrix}$$

$$\text{the transpose of the cofactors} = \mathbf{A}^{*\mathrm{T}} = \begin{pmatrix} -3 & -4 \\ -2 & 5 \end{pmatrix}$$

$$\text{the inverse of } \mathbf{A} \text{ is } \mathbf{A}^{-1} = \frac{\mathbf{A}^{*\mathrm{T}}}{|\mathbf{A}|} = \frac{\begin{pmatrix} -3 & -4 \\ -2 & 5 \end{pmatrix}}{5 \times -3 - 2 \times 4}$$

$$= -\frac{1}{23} \begin{pmatrix} -3 & -4 \\ -2 & 5 \end{pmatrix}$$

$$= \begin{pmatrix} \frac{3}{23} & \frac{4}{23} \\ \frac{2}{23} & -\frac{5}{23} \end{pmatrix}$$

$$\mathbf{A}^{-1} = \begin{pmatrix} \frac{3}{23} & \frac{4}{23} \\ \frac{2}{23} & -\frac{5}{23} \end{pmatrix}$$

Pre-multiplying each side of (1) by $\mathbf{A}^{-1}$

$$\begin{pmatrix} \frac{3}{23} & \frac{4}{23} \\ \frac{2}{23} & -\frac{5}{23} \end{pmatrix} \begin{pmatrix} 5 & 4 \\ 2 & -3 \end{pmatrix} \begin{pmatrix} x \\ y \end{pmatrix} = \begin{pmatrix} \frac{3}{23} & \frac{4}{23} \\ \frac{2}{23} & -\frac{5}{23} \end{pmatrix} \begin{pmatrix} -2 \\ 13 \end{pmatrix} = \begin{pmatrix} \frac{46}{23} \\ -\frac{69}{23} \end{pmatrix}$$

$$\mathrm{I} \begin{pmatrix} x \\ y \end{pmatrix} = \begin{pmatrix} \frac{46}{23} \\ -\frac{69}{23} \end{pmatrix} = \begin{pmatrix} 2 \\ -3 \end{pmatrix} \Rightarrow x = 2 \text{ and } y = -3.$$

Examination Test Papers.

GCE Advanced Level.

Further Pure Mathematics FP1

Test Paper 7

Time: 1 hour and 30 minutes.

Instructions and Information

Candidates may use any calculator allowed by the regulations of their Examination Board.

Full marks are awarded for correct answers to ALL questions.

This paper has nine questions.

You can start working with any question and you must label clearly all parts.

1. Find (i) the moduli and (ii) the arguments of the complex roots for

$$z^2 + 2z + 2 = 0.$$ **(6)**

---

2. Find the sum and the difference of two complex numbers $z_1 = 2 + 5i$ and $z_2 = 3 + 2i$

(i) algebraically

(ii) graphically. **(10)**

---

3. Sketch the curve $y = \dfrac{1}{1-x}$ by considering the gradient and stating the asymptotes. **(8)**

---

4. The cubic function $f(x) = 2x^3 - 3x^2 - 11x + 8$ was drawn and the positive roots were found to be 0.68 and 2.93 to two decimal places when $f(x) = 0$.

Use the Newton-Raphson method once in order to find improved roots.

Give your answers to 3 s.f. **(10)**

---

5. A triangle has the following vertices $A(0, 1)$, $B(1, 0)$ and $C(3, 3)$.

Find the coordinates of the triangle:

(a) Reflected in the x-axis.

(b) Reflected in the y-axis.

(c) Rotated about the origin through 90° in a clockwise direction.

(d) Reduced by the transformation $= \begin{pmatrix} \frac{1}{2} & 0 \\ 0 & \frac{1}{2} \end{pmatrix}$.

(e) Enlarged by the transformation $\begin{pmatrix} 2 & 0 \\ 0 & 2 \end{pmatrix}$. **(10)**

---

6. The roots of the equation $2x^2 + 6x + 3 = 0$ are α and β.

Find the quadratic equation whose roots are $\dfrac{1}{\alpha}$ and $\dfrac{1}{\beta}$. **(6)**

---

7. Use standard formulae to find expressions for

(i) $\sum_{r=1}^{n}(r+1)(r+3)$

(ii) $\sum_{r=1}^{n} r^2(r+1)$

(iii) $\sum_{r=1}^{n}(r+1)(r+5)$. **(14)**

---

8. Prove that $\left(\frac{z_1}{z_2}\right)^* = \frac{z_1^*}{z_2^*} =$ where $z_1 = a_1 + b_1 i$ and $z_2 = a_2 + b_2 i$

where $(a_1, b_1, a_2, b_2 \in \mathbb{R})$. **(5)**

---

9. Sketch the graphs and the directrices:

(i) $y^2 = 4ax$ (ii) $y^2 = -4ax$ (iii) $x^2 = 4ay$ (iv) $x^2 = -4ay$. **(6)**

TOTAL FOR PAPER: 75 MARKS

Examination Test Papers.

GCE Advanced Level.

Test Paper 7 Solutions

Further Pure Mathematics FP1

1. $z^2 + 2z + 2 = 0$

$$z = \frac{-2 \pm \sqrt{4-8}}{2} = -1 \pm i$$

$z_1 = -1 + i$ or $z_2 = -1 - i$

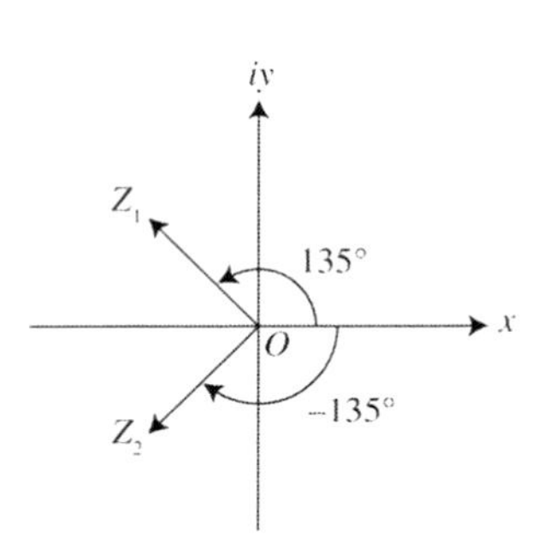

$$|z_1| = \sqrt{(-1)^2 + 1^2} = \sqrt{2}$$

$$|z_2| = \sqrt{(-1)^2 + (-1)^2} = \sqrt{2}$$

$$\arg z_1 = 135° = \frac{3\pi}{4}$$

$$\arg z_2 = -135° = -\frac{3\pi}{4}.$$

---

2. (i) $z_1 + z_2 = 2 + 5i + 3 + 2i = 5 + 7i$

$z_1 - z_2 = (2 + 5i) - (3 + 2i) = -1 + 3i.$

The real terms are added or subtracted and the imaginary terms are added or subtracted separately.

(ii)

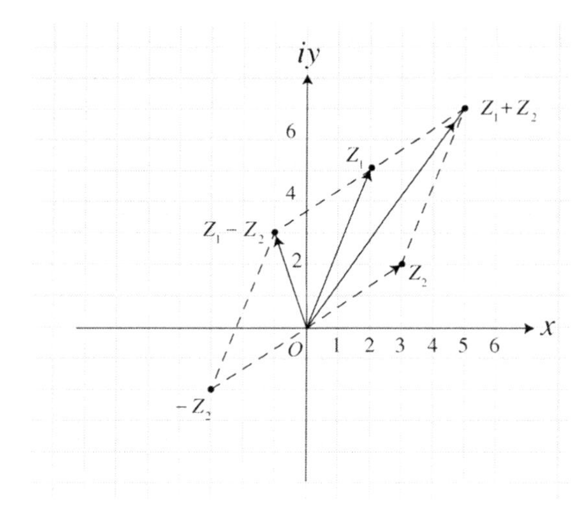

---

3. $y = \dfrac{1}{1-x} = (1-x)^{-1}$

$$\frac{dy}{dx} = (-1)(-1)(1-x)^{-2} = \frac{1}{(x-2)^2}.$$

The gradient is positive at any point of x.

Asymptotes are $\boxed{x = 1}$ and $\boxed{y = 0}$

$y \to \infty \quad$ when $x \to 1$

when $x \to \infty \quad y \to 0.$

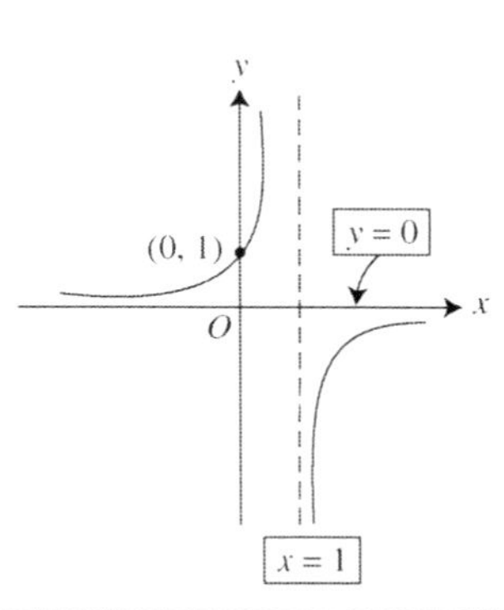

---

4. $\quad f(x) = 2x^3 - 3x^2 - 11x + 8$

$f(0.68) = 2 \times 0.68^3 - 3 \times 0.68^2 - 11 \times 0.68 + 8 = -0.238336$

$f'(x) = 6x^2 - 6x - 11$

$f'(0.68) = 6 \times 0.68^2 - 6 \times 0.68 - 11 = -12.3056$

$$x_{n+1} = x_n - \frac{f(x_n)}{f'(x_n)} = 0.68 - \frac{(-0.238336)}{(-12.3056)} = 0.68 - 0.019368092 = 0.660631907$$

$x_{n+1} = 0.660631907$

$x_{n+1} = 0.661$

$f(2.93) = 2 \times 2.93^3 - 3 \times 2.93^2 - 11 \times 2.93 + 8 = 0.322814$

$f'(2.93) = 6 \times (2.93)^2 - 6 \times 2.93 - 11 = 22.9294$

$$x_{n+1} = 2.93 - \frac{0.322814}{22.9294} = 2.9159214 = 2.92 \text{ to 3 s.f.}$$

---

5. a) $\begin{pmatrix} 1 & 0 \\ 0 & -1 \end{pmatrix} \begin{pmatrix} 0 & 1 & 3 \\ 1 & 0 & 3 \end{pmatrix} = \begin{pmatrix} 0 & 1 & 3 \\ -1 & 0 & -3 \end{pmatrix}.$

b) $\begin{pmatrix} -1 & 0 \\ 0 & 1 \end{pmatrix} \begin{pmatrix} 0 & 1 & 3 \\ 1 & 0 & 3 \end{pmatrix} = \begin{pmatrix} 0 & -1 & -3 \\ 1 & 0 & 3 \end{pmatrix}.$

c) $\begin{pmatrix} 0 & 1 \\ -1 & 0 \end{pmatrix} \begin{pmatrix} 0 & 1 & 3 \\ 1 & 0 & 3 \end{pmatrix} = \begin{pmatrix} 1 & 0 & 3 \\ 0 & -1 & -3 \end{pmatrix}.$

d) $\begin{pmatrix} \frac{1}{2} & 0 \\ 0 & \frac{1}{2} \end{pmatrix} \begin{pmatrix} 0 & 1 & 3 \\ 1 & 0 & 3 \end{pmatrix} = \begin{pmatrix} 0 & \frac{1}{2} & \frac{3}{2} \\ \frac{1}{2} & 0 & \frac{3}{2} \end{pmatrix}.$

e) $\begin{pmatrix} 2 & 0 \\ 0 & 2 \end{pmatrix} \begin{pmatrix} 0 & 1 & 3 \\ 1 & 0 & 3 \end{pmatrix} = \begin{pmatrix} 0 & 2 & 6 \\ 2 & 0 & 6 \end{pmatrix}.$

---

6. $2x^2 + 6x + 3 = 0$

$$\alpha + \beta = -\frac{6}{2} = -3 \quad \text{and} \quad \alpha\beta = \frac{3}{2}$$

$$x^2 - \left(\frac{1}{\alpha} + \frac{1}{\beta}\right)x + \frac{1}{\alpha} \times \frac{1}{\beta} = 0$$

$$x^2 - \frac{\alpha + \beta}{\alpha\beta}x + \frac{1}{\alpha\beta} = 0$$

$$\alpha\beta x^2 - (\alpha + \beta)x + 1 = 0 \qquad \boxed{3x^2 + 6x + 2 = 0}$$

---

7. (i) $$\sum_{r=1}^{n}(r+1)(r+3) = \sum_{r=1}^{n}(r^2 + 4r + 3) = \sum_{r=1}^{n} r^2 + 4\sum_{r=1}^{n} r + 3\sum_{r=1}^{n} 1$$

$$= \frac{n(n+1)(2n+1)}{6} + \frac{4n(n+1)}{2} + 3n$$

$$= \frac{n(n+1)(2n+1) + 12n(n+1) + 18n}{6}$$

$$= \frac{n}{6}\left[(n+1)(2n+1) + 12(n+1) + 18\right]$$

$$= \frac{n}{6}\left(2n^2 + 2n + n + 1 + 12n + 12 + 18\right)$$

$$\boxed{\sum_{r=1}^{n}(r+1)(r+3) = \frac{n}{6}(2n^2 + 15n + 31)}$$

(ii) $$\sum_{r=1}^{n} r^2(r+1) = \sum_{r=1}^{n} r^3 + \sum_{r=1}^{n} r^2$$

$$= \frac{n^2(n+1)^2}{4} + \frac{n(n+1)(2n+1)}{6}$$

$$= \frac{3n^2(n+1)^2 + 2n(n+1)(2n+1)}{12}$$

$$= \frac{n(n+1)}{12}\left[3n(n+1) + 4n + 2\right] = \frac{n(n+1)}{12}\left(3n^2 + 3n + 4n + 2\right)$$

$$\sum_{r=1}^{n} r^2(r+1) = \frac{n(n+1)}{12}\left(3n^2 + 7n + 2\right).$$

(iii) $\sum_{r=1}^{n}(r+1)(r+5) = \sum_{r=1}^{n}(r^2+6r+5) = \sum_{r=1}^{n} r^2 + 6\sum_{r=1}^{n} r + 5\sum_{r=1}^{n} 1$

$$= \frac{n(n+1)(2n+1)}{6} + \frac{6n(n+1)}{2} + 5n$$

$$= \frac{n(n+1)(2n+1) + 18n(n+1) + 30n}{6}$$

$$= \frac{n}{6}\left[(n+1)(2n+1) + 18n + 18 + 30)\right]$$

$$= \frac{n}{6}(2n^2 + 3n + 1 + 18n + 48) = \frac{n}{6}(2n^2 + 21n + 49)$$

$$\therefore \sum_{r=1}^{n}(r+1)(r+5) = \frac{n}{6}(n+7)(2n+7).$$

---

8. $\left(\frac{a_1 + b_1 i}{a_2 + b_2 i}\right)^* = \left[\frac{(a_1 + b_1 i)}{(a_2 + b_2 i)} \times \frac{(a_2 - b_2 i)}{(a_2 + b_2 i)}\right]^* = \left[\frac{a_1 a_2 - a_1 b_2 i + a_2 b_1 i + b_1 b_2}{{a_2}^2 + {b_2}^2}\right]^*$

$$= \frac{a_1 a_2 + b_1 b_2}{{a_2}^2 + {b_2}^2} - \frac{a_2 b_1 - a_1 b_2}{{a_2}^2 + {b_2}^2} i$$

$$\frac{{z_1}^*}{{z_2}^*} = \frac{(a_1 - b_1 i)}{(a_2 + b_2 i)} \times \frac{(a_2 + b_2 i)}{(a_2 + b_2 i)} = \frac{a_1 a_2 - b_1 a_2 i}{{a_2}^2 + {b_2}^2} + \frac{a_1 b_2 i + b_1 b_2}{{a_2}^2 + {b_2}^2}$$

$$\therefore \left(\frac{z_1}{z_2}\right)^* = \frac{{z_1}^*}{{z_2}^*}.$$

9.

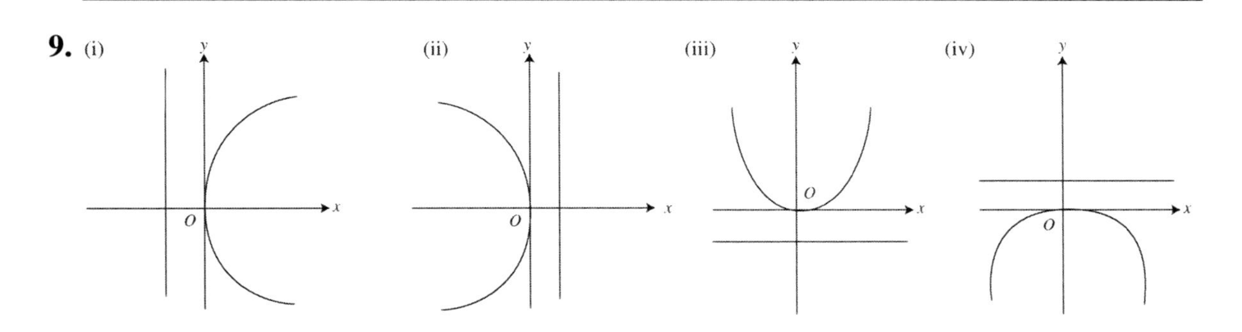

Examination Test Papers.

GCE Advanced Level.

Further Pure Mathematics FP1

Test Paper 8

Time: 1 hour and 30 minutes.

Instructions and Information

Candidates may use any calculator allowed by the regulations of their Examination Board.

Full marks are awarded for correct answers to ALL questions.

This paper has nine questions.

You can start working with any question and you must label clearly all parts.

1. The complex numbers z_1 and z_2 are given

$$z_1 = 3 + 4i$$
$$z_2 = 5 + 12i.$$

(a) Find $\frac{z_1}{z_2}$ in the form $x + yi$ where $(x, y \in \mathbb{R})$. **(3)**

(b) Find $\arg\left(\frac{z_1}{z_2}\right)$. **(5)**

---

2. The vertices of a triangle T are given by $A(-3, 3)$, $B(0, 5)$ and $C(2, 4)$.

Determine the vertices of a triangle T' of the reflected triangle in the $x-$axis. **(4)**

Draw these triangles. **(2)**

State the four properties of the reflections **(2)**

---

3. The roots of the equation $x^2 - 4x + 2 = 0$ are α and β.

(a) show that $\alpha^3 + \beta^3 = 40$. **(4)**

(b) The roots of the equation $x^2 - px + q = 0$ are $2\beta + \alpha$ and $2\alpha + \beta$,

calculate the value of p and the value of q. **(6)**

---

4. Evaluate (i) $\sum_{r=1}^{25}(r+1)(r+2)$ **(6)**

(ii) $\sum_{r=1}^{10}(r^2 + 2^r)$ **(3)**

(iii) $\sum_{r=1}^{20} r(3r - 5)$. **(3)**

---

5. (a) Show that $f(x) = x^3 - 2x - 5x = 0$ has a root α that lies in the interval [2, 2.1].

Find this root using linear interpolation correct to one decimal place.

(b) Use Newton-Raphson formula to find an improved root to five decimal places. **(8)**

---

6. Show that the line $y = mx - 5m^2$ is a tangent to the parabola $x^2 = 20y$. **(4)**

---

7. a) If $\mathbf{A} = \begin{pmatrix} 2 & 1 \\ -1 & 3 \end{pmatrix}$ and $\mathbf{B} = \begin{pmatrix} -1 & 2 \\ 1 & -2 \end{pmatrix}$.

Determine (i) **AB** and (ii) **BA**.

Explain that the product is non-commutative. **(4)**

b) Find $\mathbf{A}^{-1}$ and $\mathbf{B}^{-1}$ if possible. **(6)**

---

8. Show that $$\sum_{r=1}^{n} \frac{1}{(r+2)(r+4)} = \frac{7}{24} - \frac{1}{2}\frac{2n+7}{(n+3)(n+4)}.$$ **(8)**

---

9. (a) Find the product $\begin{pmatrix} \cos\theta & \sin\theta \\ -\sin\theta & \cos\theta \end{pmatrix}\begin{pmatrix} x \\ y \end{pmatrix}$.

(b) Find the product $\begin{pmatrix} \cos\theta & -\sin\theta \\ \sin\theta & \cos\theta \end{pmatrix}\begin{pmatrix} x \\ y \end{pmatrix}$.

State the transformations incurred in (a) and (b). **(7)**

TOTAL FOR PAPER: 75 MARKS

Examination Test Papers.

GCE Advanced Level.

Test Paper 8 Solutions

Further Pure Mathematics FP1

1. (a) $\dfrac{z_1}{z_2} = \dfrac{3+4i}{5+12i} \times \dfrac{5-12i}{5-12i} = \dfrac{15+20i-36i+48}{25+144} = \dfrac{63}{169} - \dfrac{16}{169}i$

$x = \dfrac{63}{169}$ and $y = -\dfrac{16}{169}$

(b) $\arg\left(\dfrac{z_1}{z_2}\right) = \arg\left(\dfrac{63}{169} - \dfrac{16}{169}i\right) = \theta$

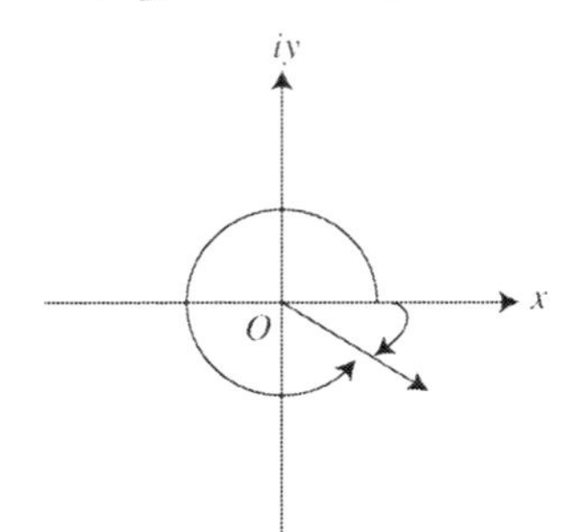

$\tan\theta = -\dfrac{16}{63} \Rightarrow \theta = -14.3°$ to 3 s.f.

or $360° - 14.3° = 345.7° = 346°$ to 3 s.f.

---

2. The position vectors of the vertices required are found by pre-multiplying the matrix

$\begin{pmatrix} -3 & 0 & 2 \\ 3 & 5 & 4 \end{pmatrix}$ by $\begin{pmatrix} 1 & 0 \\ 0 & -1 \end{pmatrix}$.

A B C

$$\begin{pmatrix} 1 & 0 \\ 0 & -1 \end{pmatrix}\begin{pmatrix} -3 & 0 & 2 \\ 3 & 5 & 4 \end{pmatrix} = \begin{pmatrix} -3 & 0 & 2 \\ -3 & -5 & -4 \end{pmatrix}.$$

A B C   A′ B′ C′

2×2 2×3 2×3

The coordinates of the reflected triangle T' in the x-axis are

A′ $(-3, -3)$, B′$(0, -5)$ and C′ $(2, -4)$.

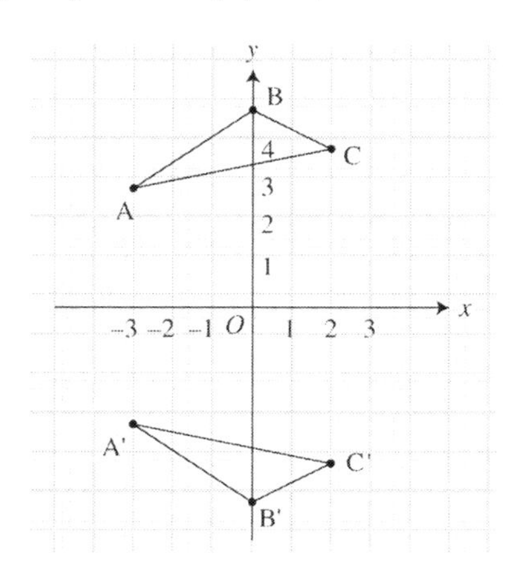

Properties of reflections.

a) The size and shape of the object is the same as the size and shape of the image and vice versa.

b) The distance of the point object from the mirror line is the same as the distance of the point image from the mirror.

c) The mirror line bisects the angle between a line and its image.

d) Reflection is its own inverse A$'$ is the image of A, A is the image A$'$.

---

3. (a) $\alpha^3 + \beta^3 = (\alpha + \beta)^3 - 3\alpha^2\beta - 3\alpha\beta^2$

but $\alpha + \beta = -\dfrac{(-4)}{1} = 4 \qquad \alpha\beta = \dfrac{2}{1} = 2$

$$\alpha^3 + \beta^3 = (4)^3 - 3\alpha\beta(\alpha + \beta)$$

$$= 64 - 3 \times 2 \times 4 = 64 - 24$$

$$= 40.$$

(b) $2\beta + \alpha + 2\alpha + \beta = -\dfrac{-p}{1} = p = 2(\alpha + \beta) + \alpha + \beta = 3(\alpha + \beta) = 3 \times 4 = 12$

$$(2\beta + \alpha)(2\alpha + \beta) = \frac{q}{1} = q$$

$$4\alpha\beta + 2\beta^2 + 2\alpha^2 + \alpha\beta = 5\alpha\beta + 2(\alpha^2 + \beta^2) = 5\alpha\beta + 2\big[(\alpha + \beta)^2 - 2\alpha\beta\big] = q$$

$$q = 5\alpha\beta + 2\big[(\alpha + \beta)^2 - 2\alpha\beta\big] = 5 \times 2 + 2\big[4^2 - 2 \times 2\big] = 10 + 2 \times 12 = 34$$

$\therefore \boxed{p = 12} \qquad \boxed{q = 34}$.

---

4. (i) $\displaystyle\sum_{r=1}^{n}(r + 1)(r + 2) = \sum_{r=1}^{n}(r^2 + 3r + 2)$

$$= \sum_{r=1}^{n} r^2 + 3\sum_{r=1}^{n} r + 2\sum_{r=1}^{n} 1$$

$$= \frac{n(n + 1)(2n + 1)}{6} + \frac{3}{2}n(n + 1) + 2n$$

$$\therefore \sum_{r=1}^{n}(r+1)(r+2) = \frac{n}{6}\left[(n+1)(2n+1)+9(n+1)+12\right]$$

$$= \frac{n}{6}\left(2n^2+3n+1+9n+9+12\right)$$

$$= \frac{n}{6}\left(2n^2+12n+22\right)$$

$$= \frac{n}{3}\left(n^2+6n+11\right)$$

$$\sum_{r=1}^{25}(r+1)(r+2) = \frac{25}{3}\left(625+150+11\right) = \frac{25}{3} \times 786 = 25 \times 262 = 6550$$

(ii) $$\sum_{r=1}^{10}(r^2+2^r) = \sum_{r=1}^{10} r^2 + \sum_{r=1}^{10} 2^r$$

$$= \frac{10 \times 11 \times 21}{6} + \left(2+2^2+2^3+\cdots+2^{10}\right)$$

$$= 5 \times 11 \times 7 + 2046 = 385 + 2046 = 2431$$

(iii) $$\sum_{r=1}^{20} r(3r-5) = 3\sum_{r=1}^{20} r^2 - 5\sum_{r=1}^{20} r$$

$$= 3 \times \frac{20 \times 21 \times 41}{6} - 5 \times \frac{20 \times 21}{2}$$

$$= 210 \times 41 - 50 \times 21 = 8610 - 1050 = 7560.$$

---

5. (a) $f(x) = x^3 - 2x - 5 = 0$

$f(2) = 2^3 - 2 \times 2 - 5 = 8 - 9 = -1$

$f(2.1) = 2.1^3 - 2 \times 2.1 - 5 = 0.061.$

There is a change of sign between 2 and 2.1 and therefore there is an approximate solution in this interval.

(b) $$x_{n+1} = x_n - \frac{f(x_n)}{f'(x_n)}$$

$f(2.1) = 0.061$.

Using linear interpolation $\dfrac{0.061}{1} = \dfrac{2.1 - x}{x - 2.1} \Rightarrow x = 2.1$.

$$f'(x) = 3x^2 - 2 \Rightarrow f'(2.1) = 3(2.1)^2 - 2 = 11.23$$

$$x_{n+1} = 0.061 - \frac{f(x_n)}{f'(x_n)} = 0.061 - \frac{0.061}{11.23} = 0.061 - 5.431878896 \times 10^{-3}$$

$$= 0.055568121 = 0.05557 \text{ to 5 d.p.}$$

---

6. $x^2 = 20y$

$x^2 = (mx - 5m^2)20 \Rightarrow x^2 - 20mx + 100m^2 = 0$.

The discriminant $= D = b^2 - 4ac = 400m^2 - 4 \times 1 \times 100m^2 = 0$.

Therefore the line touches the parabola.

---

7. (a) (i) $\mathbf{AB} = \begin{pmatrix} 2 & 1 \\ -1 & 3 \end{pmatrix} \begin{pmatrix} -1 & 2 \\ 1 & -2 \end{pmatrix}$

$$= \begin{pmatrix} -2+1 & 4-2 \\ 1+3 & -2-6 \end{pmatrix} = \begin{pmatrix} -1 & 2 \\ 4 & -8 \end{pmatrix}$$

(ii) $\mathbf{BA} = \begin{pmatrix} -1 & 2 \\ 1 & -2 \end{pmatrix} \begin{pmatrix} 2 & 1 \\ -1 & 3 \end{pmatrix}$

$$= \begin{pmatrix} -2-2 & -1+6 \\ 2+2 & 1-6 \end{pmatrix} = \begin{pmatrix} -4 & 5 \\ 4 & -5 \end{pmatrix}$$

$\mathbf{AB} \neq \mathbf{BA}$ non commutative.

(b) $\mathbf{A} = \begin{pmatrix} 2 & 1 \\ -1 & 3 \end{pmatrix}$

minors of $\mathbf{A} = \begin{pmatrix} 3 & -1 \\ 1 & 2 \end{pmatrix}$

cofactor of $\mathbf{A} = \mathbf{A}^* = \begin{pmatrix} 3 & 1 \\ -1 & 2 \end{pmatrix}$

$$\mathbf{A}^{*\mathrm{T}} = \begin{pmatrix} 3 & -1 \\ 1 & 2 \end{pmatrix}$$

$$\mathbf{A}^{-1} = \frac{\mathbf{A}^{*\mathrm{T}}}{|\mathbf{A}|} = \frac{\begin{pmatrix} 3 & -1 \\ 1 & 1 \end{pmatrix}}{7} = \begin{pmatrix} \frac{3}{7} & -\frac{1}{7} \\ \frac{1}{7} & \frac{2}{7} \end{pmatrix}$$

$$\mathbf{A}^{-1}\mathbf{A} = \mathbf{A}\mathbf{A}^{-1} = \begin{pmatrix} 1 & 0 \\ 0 & 1 \end{pmatrix} = \text{unit matrix}$$

$$\begin{pmatrix} \frac{3}{7} & -\frac{1}{7} \\ \frac{1}{7} & \frac{2}{7} \end{pmatrix} \begin{pmatrix} 2 & 1 \\ -1 & 3 \end{pmatrix} = \begin{pmatrix} \frac{6}{7} + \frac{1}{7} & \frac{3}{7} - \frac{3}{7} \\ \frac{2}{7} - \frac{2}{7} & \frac{1}{7} + \frac{6}{7} \end{pmatrix} = \begin{pmatrix} 1 & 0 \\ 0 & 1 \end{pmatrix} = \mathbf{I}$$

$$\mathbf{B}^{-1} = \frac{\mathbf{B}^{*\mathrm{T}}}{|\mathbf{B}|} = \frac{\begin{pmatrix} -2 & -2 \\ -1 & -1 \end{pmatrix}}{0} \text{ not defined } |\mathbf{B}| = \begin{vmatrix} -1 & 2 \\ 1 & -2 \end{vmatrix} = \mathbf{2} - \mathbf{2} = \mathbf{0}$$

$$\mathbf{B}^{*} = \begin{pmatrix} -2 & -1 \\ -2 & -1 \end{pmatrix} \text{ and } \mathbf{B}^{*\mathrm{T}} = \begin{pmatrix} -2 & -2 \\ -1 & -1 \end{pmatrix} \qquad \text{it is singular.}$$

---

8. $$\sum_{r=1}^{n} \frac{1}{(r+2)(r+4)} \qquad \ldots (1)$$

$$\frac{1}{(r+2)(r+4)} \equiv \frac{A}{r+2} + \frac{B}{r+4}$$

$$1 \equiv A(r+4) + B(r+2)$$

if $r = -2$, $\quad A = \frac{1}{2}$ and if $r = -4$, $\quad B = -\frac{1}{2}$

$$\frac{1}{(r+2)(r+4)} \equiv \frac{1}{2(r+2)} - \frac{1}{2(r+4)}$$

substituting in (1)

$$\frac{1}{2}\sum_{r=1}^{n} \frac{1}{r+2} - \frac{1}{2}\sum_{r=1}^{n} \frac{1}{r+4} = \frac{1}{2}\left(\frac{1}{3} + \frac{1}{4} + \frac{1}{5} + \ldots \frac{1}{n+2}\right)$$

$$-\frac{1}{2}\left(\frac{1}{5} + \frac{1}{6} + \cdots + \frac{1}{n+4}\right)$$

let $S = \frac{1}{5} + \frac{1}{6} + \cdots + \frac{1}{n+2}$

$$\sum_{r=1}^{n} \frac{1}{(r+2)(r+4)} = \frac{1}{2}\left(\frac{1}{3} + \frac{1}{4} + S\right) - \frac{1}{2}\left(S + \frac{1}{n+3} + \frac{1}{n+4}\right)$$

$$= \frac{1}{2}\left(\frac{1}{3} + \frac{1}{4}\right) + \frac{1}{2}S - \frac{1}{2}S - \frac{1}{2(n+3)} - \frac{1}{2(n+4)}$$

$$= \frac{7}{24} - \frac{1}{2(n+3)} - \frac{1}{2(n+4)}$$

$$= \frac{7}{24} - \frac{1}{2}\frac{2n+7}{(n+3)(n+4)}.$$

---

9. (a) $\begin{pmatrix} \cos\theta & \sin\theta \\ -\sin\theta & \cos\theta \end{pmatrix} \begin{pmatrix} x \\ y \end{pmatrix} = \begin{pmatrix} x\cos\theta + y\sin\theta \\ -x\sin\theta + y\cos\theta \end{pmatrix}.$

$2 \times 2 \qquad 2 \times 1 \qquad 2 \times 1$

A rotation about the origin in a clockwise direction.

(b) $\begin{pmatrix} \cos\theta & -\sin\theta \\ \sin\theta & \cos\theta \end{pmatrix} \begin{pmatrix} x \\ y \end{pmatrix} = \begin{pmatrix} x\cos\theta - y\sin\theta \\ x\sin\theta + y\cos\theta \end{pmatrix}.$

$2 \times 2 \qquad 2 \times 1 \qquad 2 \times 1$

A rotation about the origin in an anticlockwise direction.

Examination Test Papers.

GCE Advanced Level.

Further Pure Mathematics FP1

Test Paper 9

Time: 1 hour and 30 minutes.

Instructions and Information

Candidates may use any calculator allowed by the regulations of their Examination Board.

Full marks are awarded for correct answers to ALL questions.

This paper has nine questions.

You can start working with any question and you must label clearly all parts.

1. Given that $z_1 = 3 - 4i$ and $z_2 = \dfrac{5 + 12i}{z_1}$.

(a) Find z_2 in the form $a + ib$, where $a, b \in \mathbb{R}$.

(b) Show on an Argand diagram the points P and Q which are representing z_1 and z_2. **(6)**

---

2. If α and β are the roots of the quadratic equation $x^2 - 4x + 5 = 0$.

Find the quadratic equation which has roots α^3 and β^3. **(6)**

---

3. Find a polynomial expression for $\displaystyle\sum_{r=1}^{n} r^2(r + 2)$.

Hence evaluate $\displaystyle\sum_{r=1}^{20} r^2(r + 2)$. **(8)**

---

4. The quadratic function is shown in the figure.

Find the equation of this function.

What is the value of c?

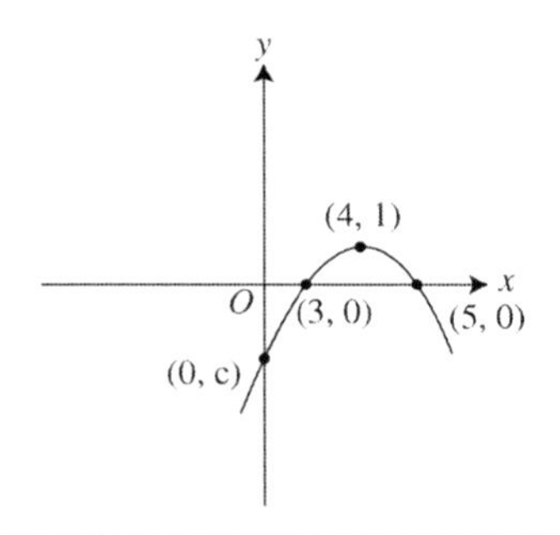

(10)

---

5. Sketch the line $x + y = 5$ and the exponential expression $y = 3^x$ on the same graph, in the first quadrant. Let $P(x, y)$ be the intersection of these graphs. Find the interval values of x that makes $f(x) = 0$. Use interval bisection to find an approximate root of $f(x) = 0$ to 3 s.f. **(10)**

---

6. $f(x) = 3^x + x - 5$.

Use a linear interpolation to find an approxinate solution for $f(x) = 0$, in the interval $[1, 2]$ to 3 decimal places **(12)**

---

7. Use Newton-Raphson formula for $f(x) = 3^x + x - 5$ to improve the approximate root for $f(x) = 0$ if $x = 1.125$ to 3 d.p. **(8)**

---

8. Use the method of mathematical induction to show that $\sum_{r=1}^{n} r^3 = \frac{n^2(n+1)^2}{4}$. **(7)**

---

9. Solve the linear equations simultaneously using matrices.

$$5x - 7y = 46$$
$$3x + 4y = 3.$$

(8)

TOTAL FOR PAPER: 75 MARKS

Examination Test Papers.

GCE Advanced Level.

Test Paper 9 Solutions

Further Pure Mathematics FP1

1. (a) $z_2 = \frac{(5+12i)}{(3-4i)} \times \frac{(3+4i)}{(3+4i)} = \frac{15+36i+20i-48}{25} = -\frac{33}{25} + \frac{56}{25}i.$

(b) $|z_1| = 5 \qquad |z_2| = \sqrt{\left(-\frac{33}{25}\right)^2 + \left(\frac{56}{25}\right)^2} = 2.6$

$\arg z_1 = -53.1°$

$\arg z_2 = 120.5°$

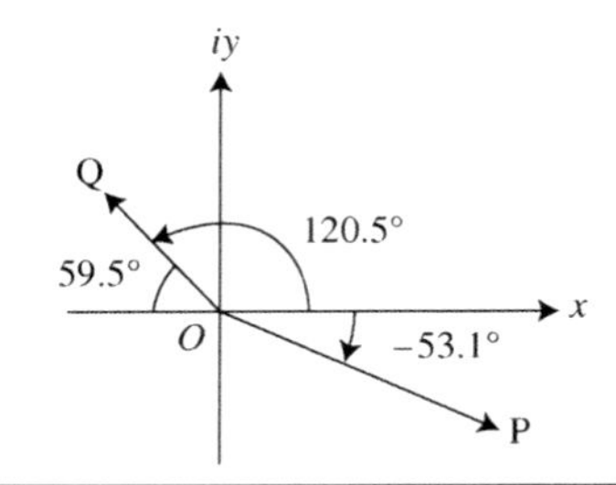

---

2. $x^2 - 4x + 5 = 0$

$\alpha + \beta = 4 \qquad \alpha\beta = 5$

$x^2 - (\alpha^3 + \beta^3)x + \alpha^3\beta^3 = 0$

$\alpha^3 + \beta^3 = (\alpha+\beta)^3 - 3\alpha^2\beta - 3\alpha\beta^2$

$\qquad = (\alpha+\beta)^3 - 3\alpha\beta(\alpha+\beta) = 4^3 - 3 \times 5 \times 4 = 64 - 60 = 4$

$x^2 - 4x + \ 5^3 = 0$

$x^2 - 4x + 125 = 0.$

---

3. $\sum_{r=1}^{n} r^2(r+2) = \sum_{r=1}^{n} r^3 + 2\sum_{r=1}^{n} r^2$

$$= \left[\frac{n(n+1)}{2}\right]^2 + 2 \times \frac{n(n+1)(2n+1)}{6}$$

$$= \frac{n^2(n+1)^2}{4} + \frac{n(n+1)(2n+1)}{3}$$

$$= \frac{3n^2(n+1)^2 + 4(n+1)(2n+1)}{12}$$

$$= \frac{(n+1)}{12}\left[3n^2(n+1) + 4(2n+1)\right]$$

$$= \frac{(n+1)}{12}(3n^3 + 3n^2 + 8n + 4)$$

$$\sum_{r=1}^{20} r^2(r+2) = \frac{21}{12}\left(3 \times 20^3 + 3 \times 20^2 + 8 \times 20 + 4\right)$$

$$= \frac{7}{4}(24000 + 1200 + 320 + 4) = 7\left(6000 + 300 + 80 + 1\right)$$

$$= 7 \times 6381 = 44667.$$

---

4. The curve has a maximum and $a < 0$ if the quadratic function is

$ax^2 + bx + c = \mathrm{f}(x)$, $b^2 - 4ac > 0$

$$\mathrm{f}(x) = a\left[x^2 + \frac{b}{a}x + \frac{c}{a}\right]$$

$$= a\left[\left(x + \frac{b}{2a}\right)^2 - \frac{b^2}{4a^2} + \frac{c}{a}\right]$$

$$\mathrm{f}\left(-\frac{b}{2a}\right) = a\left(-\frac{b^2}{4a^2} + \frac{c}{a}\right) = -\frac{b^2}{4a} + c$$

$$\mathrm{f}(4) = 1$$

$$-\frac{b^2}{4a} + c = 1$$

$$4 = -\frac{b}{2a} \Rightarrow b = -8a$$

$$\mathrm{f}(3) = 0 \qquad \mathrm{f}(5) = 0$$

$$\mathrm{f}(4) = 1 \qquad 16a + 4b + c = 1 \qquad \ldots(1)$$

$$9a + 3b + c = 0 \qquad \ldots(2)$$

$$25a + 5b + c = 0 \qquad \ldots(3)$$

$$(1) - (2) \quad \boxed{7a + b = 1} \qquad \ldots(4)$$

$$(3) - (1) \quad \boxed{9a + b = -1} \qquad \ldots(5)$$

$$7a + b = 1 \qquad \ldots(4)$$

$$9a + b = -1 \qquad \ldots(5)$$

$$(5) - (4) \qquad 2a = -2$$

$$\boxed{a = -1}$$

$$7a + b = 1$$

$$b = 1 - 7a = 1 + 7 = 8$$

$$\boxed{b = 8}$$

$$16a + 4b + c = 1$$

$$-16 + 4(8) + c = 1$$

$$c = 1 + 16 - 32$$

$$c = -15$$

$$\therefore \boxed{f(x) = -x^2 + 8x - 15}$$

$$b^2 - 4ac = 64 - 4(-1)(-15) = 64 - 60 = 4$$

$$\therefore D > 0$$

5.

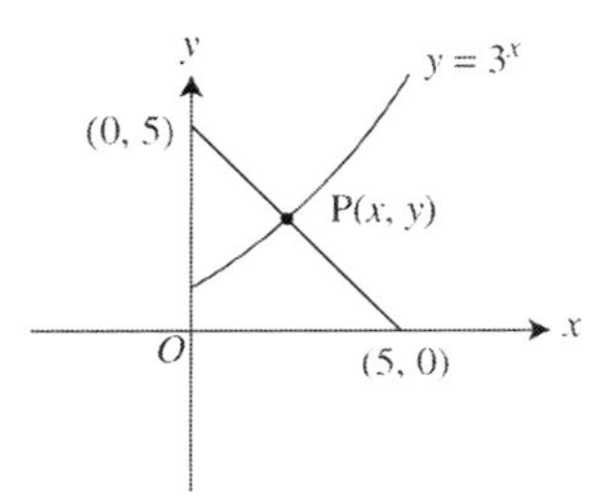

$$3^x = 5 - x$$

$$f(x) = 3^x + x - 5 = 0$$

$$f(2) = 9 + 2 - 5 = 6$$

$$f(1) = 3 + 1 - 5 = -1.$$

The approximate solution is in the interval values [1, 2].

$$\frac{1+2}{2} = 1.5 \qquad \text{mid-point}$$

$$f(1.5) = 3^{1.5} + 1.5 - 5 = 3^{1.5} - 3.5 = 5.196152423 - 3.5$$

$$= 1.696152423$$

$$\frac{1+1.5}{2} = 1.25 \qquad \text{mid-point}$$

$$f(1.25) = 3^{1.25} + 1.25 - 5 = 0.198222038$$

$$\frac{1.25 + 1.5}{2} = \frac{2.75}{2} = 1.375 \quad \text{mid-point this would not give a sign change}$$

$$\frac{1 + 1.25}{2} = \frac{2.25}{2} = 1.125$$

$$f(1.125) = 3^{1.125} + 1.125 - 5$$

$$= -0.433391928$$

$$\therefore x = 1.125.$$

6.

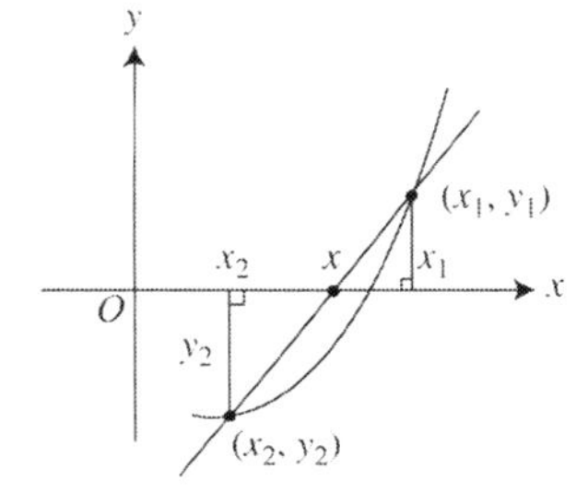

Using similar triangles

$$\frac{y_1}{y_2} = \frac{x_1 - x}{x - x_2} \qquad \ldots (1)$$

$$f(x_1) = 3^{x_1} + x_1 - 5$$

$$f(2) = 3^2 + 2 - 5 = \quad 6 = y_1 \qquad x_1 = 2$$

$$f(1) = 3^1 + 1 - 5 = -1 = y_2 \qquad x_2 = 1.$$

There is a change of sign in the interval [1, 2].

Using equation (1)

$$\frac{6}{1} = \frac{2 - x}{x - 1} \Rightarrow 6x - 6 = 2 - x$$

$$7x = 8$$

$$x = \frac{8}{7} = 1.142857143$$

$$f\,(1.142857143) = 3^{1.142857143} + 1.142857143 - 5$$

$$= -0.347350418$$

$x_1 = 1.142857143 \qquad y_1 = 0.347350418$

$x_2 = 1 \qquad y_2 = 1$

$$\frac{y_1}{y_2} = \frac{x_1 - x}{x - x_2}$$

$$\frac{0.347350418}{1} = \frac{1.142857143 - x}{x - 1}$$

$$0.347350418x + x = 1.142857143 + 0.347350418$$

$$x = \frac{1.489936132}{1.347350418} = 1.105826749 = 1.106.$$

---

7. $$x_{n+1} = x_n - \frac{f(x_n)}{f'x_n}$$

$$f(x) = 3^x + x - 5$$

$$f(1.125) = 3^{1.125} + 1.125 - 5 = -0.433391928$$

$$f'(x) = 3^x \ln 3 + 1$$

$$f'(1.125) = 3^{1.125} \ln 3 + 1 = 4.78099292$$

$$x_{n+1} = 1.125 - \frac{(-0.433391928)}{4.78099292} = 1.125 + 0.090648937 = 1.215648937$$

$$f(1.215648937) = 3^{1.215648937} + 1.215648937 - 5 = 0.01964748.$$

---

8. $$\sum_{r=1}^{n} r^3 = 1^3 + 2^3 + 3^3 + \cdots + n^3 = n^2 \frac{(1+n)^2}{4}.$$

For $n = 1$, $1^3 = \frac{1^2(1+1)^2}{4} = 1$ it is true.

For $n = 2$, $1^3 + 2^3 = \frac{2^2(1+2)^2}{4} = 9$ it is also true.

For $n = k$, $1^3 + 2^3 + 3^3 + \cdots + k^3 = \frac{k^2(1+k)^2}{4}$ ... (1) it would be true

Adding the $(k+1)$th term to both sides of (1)

$$1^3 + 2^3 + 3^3 + \cdots + k^3 + (k+1)^3 = \frac{k^2(1+k)^2}{4} + (k+1)^3$$

$$= \frac{(1+k)^2[k^2 + 4(k+1)]}{4} = \frac{(1+k)^2(k+2)^2}{4}$$

$\therefore$ it is true for n.

9.

$$5x - 7y = 46$$
$$3x + 4y = \ \ 3$$

$$\begin{pmatrix} 5 & -7 \\ 3 & 4 \end{pmatrix}\begin{pmatrix} x \\ y \end{pmatrix} = \begin{pmatrix} 46 \\ 3 \end{pmatrix}.$$

Let $\mathbf{A} = \begin{pmatrix} 5 & -7 \\ 3 & 4 \end{pmatrix}$

the minors of $\mathbf{A} = \begin{pmatrix} 4 & 3 \\ -7 & 5 \end{pmatrix}$

the cofactors of $\mathbf{A} = \mathbf{A}^* = \begin{pmatrix} 4 & -3 \\ 7 & 5 \end{pmatrix}$

$\mathbf{A}^{*\mathrm{T}} = \begin{pmatrix} 4 & 7 \\ -3 & 5 \end{pmatrix}$

$$\mathbf{A}^{-1} = \frac{1}{|\mathbf{A}|} \quad \mathbf{A}^{*\mathrm{T}} = \begin{pmatrix} \frac{4}{41} & \frac{7}{41} \\ -\frac{3}{41} & \frac{5}{41} \end{pmatrix}$$

$$|\mathbf{A}| = \begin{vmatrix} 5 & -7 \\ 3 & 4 \end{vmatrix} = 20 + 21 = 41$$

$$\begin{pmatrix} 5 & -7 \\ 3 & 4 \end{pmatrix} = \begin{pmatrix} \frac{4}{41} & \frac{7}{41} \\ -\frac{3}{41} & \frac{5}{41} \end{pmatrix}\begin{pmatrix} x \\ y \end{pmatrix} = \begin{pmatrix} \frac{4}{41} & \frac{7}{41} \\ -\frac{3}{41} & \frac{5}{41} \end{pmatrix}\begin{pmatrix} 46 \\ 3 \end{pmatrix}$$

$$\begin{pmatrix} \frac{20}{41} + \frac{21}{41} & \frac{35}{41} - \frac{35}{41} \\ \frac{12}{41} - \frac{12}{41} & \frac{21}{41} + \frac{20}{41} \end{pmatrix}\begin{pmatrix} x \\ y \end{pmatrix} = \begin{pmatrix} \frac{184}{41} + \frac{21}{41} \\ -\frac{138}{41} + \frac{15}{41} \end{pmatrix} = \begin{pmatrix} \frac{205}{41} \\ -\frac{123}{41} \end{pmatrix}$$

$$\begin{pmatrix} 1 & 0 \\ 0 & 1 \end{pmatrix}\begin{pmatrix} x \\ y \end{pmatrix} = \begin{pmatrix} 5 \\ -3 \end{pmatrix}$$

$$\therefore \boxed{x = 5} \qquad \boxed{y = -3}$$

Examination Test Papers.

GCE Advanced Level.

Further Pure Mathematics FP1

Test Paper 10

Time: 1 hour and 30 minutes.

Instructions and Information

Candidates may use any calculator allowed by the regulations of their Examination Board.

Full marks are awarded for correct answers to ALL questions.

This paper has nine questions.

You can start working with any question and you must label clearly all parts.

1. Find $\sum_{r=1}^{n}(r^2 - r + 1)$ given $\sum_{r=1}^{n} 1 = n$ $\quad \sum_{r=1}^{n} r = \frac{n(n+1)}{2}$

and $\sum_{r=1}^{n} r^2 = \frac{n(n+1)(2n+1)}{6}$. **(5)**

---

2. $f(x) = x^3 + 3x - 5 = 0$ has an approximate root in the interval [1, 2].

Show that there is change of sign. **(3)**

---

3. The square S has vertices A (2, 2), B (4, 2), C(4, 4) and D (4, 2).

Find the vertices of S under the transformation represented by these matrices

(a) $\begin{pmatrix} 3 & 1 \\ 2 & 2 \end{pmatrix}$ (b) $\begin{pmatrix} 1 & -1 \\ 1 & 1 \end{pmatrix}$. **(5)**

---

4. (a) Determine the complex roots of the quadratic equation $3z^2 - z + 1 = 0$.

(b) Sketch these roots in an Argand diagram. **(5)**

---

5. Find the coodinates of the point of intersection, T of the normals at the points $P(2ap, -ap^2)$ and $Q(2aq, -aq^2)$ of the parabola $x^2 = -4ay$.

Determine the locus of T if $pq = -1$. **(13)**

---

6. Find the inverse matrix of $\mathbf{A} = \begin{pmatrix} \frac{3}{5} & \frac{1}{5} \\ \frac{4}{5} & \frac{3}{5} \end{pmatrix}$.

Show that $\mathbf{AA}^{-1} = \mathbf{A}^{-1}\mathbf{A} = \mathbf{I}$. **(12)**

---

7. If $z_1 = 1 + 3i$ and $z_2 = \sqrt{3} - i$ show on an Argand diagram points representing the complex numbers:

(i) $z_1 z_2$ (ii) $z_1 + z_2$ (iii) $\frac{z_2}{z_1}$. **(9)**

---

8. Form the quadratic equation for which the sum of the roots is -7 and the sum of the cubes of the root is 125. **(8)**

---

9. The exponential function 2^x is intersected by the line.

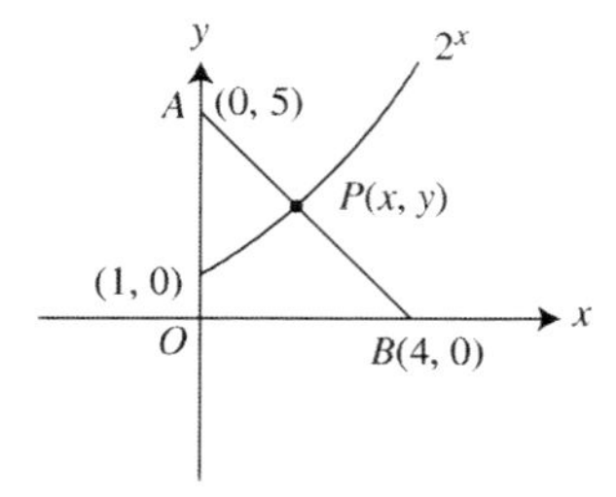

Find the coordinate values of the point of intersection x and y.

Find the interval values in which there is a change of sign.

(i) Use linear interpolation. **(5)**

(ii) Interval bisection.

(iii) Newton-Raphson formula in order to find approximate values of $f(x) = 0$ in three significant figures. **(10)**

TOTAL FOR PAPER: 75 MARKS

Examination Test Papers.

GCE Advanced Level.

Test Paper 10 Solutions

Further Pure Mathematics FP1

1. $\sum_{r=1}^{n}(r^2 - r + 1) = \sum_{r=1}^{n} r^2 - \sum_{r=1}^{n} r + \sum_{r=1}^{n} 1$

$$= \frac{n(n+1)(2n+1)}{6} - \frac{n(n+1)}{2} + n$$

$$= \frac{n}{6}\left[(n+1)(2n+1) - 3(n+1) + 6\right]$$

$$= \frac{n}{6}\left(2n^2 + 2n + n + 1 - 3n - 3 + 6\right) = \frac{n}{6}\left(2n^2 + 4\right) = \frac{n}{3}\left(n^2 + 2\right).$$

---

2. $f(x) = x^3 + 3x - 5$

$f(1) = 1 + 3 - 5 = -1$

$f(2) = 8 + 6 - 5 = 9.$ Therefore there is a change of sign .

---

3. (a) $\begin{pmatrix} 3 & 1 \\ 2 & 2 \end{pmatrix} \begin{pmatrix} 2 & 4 & 4 & 4 \\ 2 & 2 & 4 & 2 \end{pmatrix} = \begin{pmatrix} 8 & 14 & 16 & 14 \\ 8 & 12 & 16 & 12 \end{pmatrix}$

A B C D → A′ B′ C′ D′

(b) $\begin{pmatrix} 1 & -1 \\ 0 & 1 \end{pmatrix} \begin{pmatrix} 2 & 4 & 4 & 4 \\ 2 & 2 & 4 & 2 \end{pmatrix} = \begin{pmatrix} 0 & 2 & 0 & 2 \\ 4 & 6 & 8 & 6 \end{pmatrix}.$

A" B" C" D"

---

4. (a) $3z^2 - z + 1 = 0$

$$z = \frac{1 \pm \sqrt{1 - 12}}{6} = \frac{1}{6} \pm i\frac{\sqrt{11}}{6}$$

$$z_1 = \frac{1}{6} + i\frac{\sqrt{11}}{6}$$

$$z_2 = \frac{1}{6} - i\frac{\sqrt{11}}{6}.$$

(b)

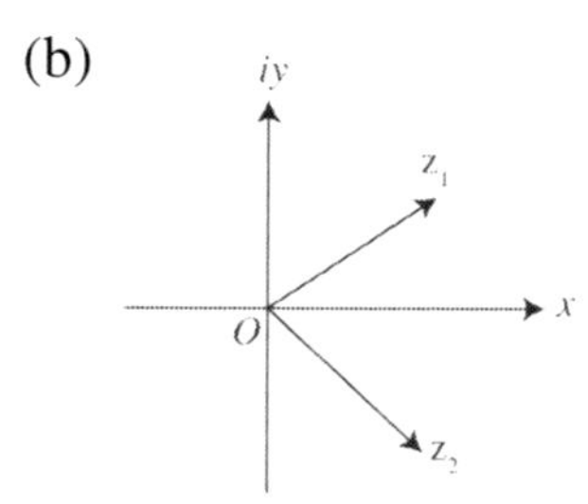

---

5. 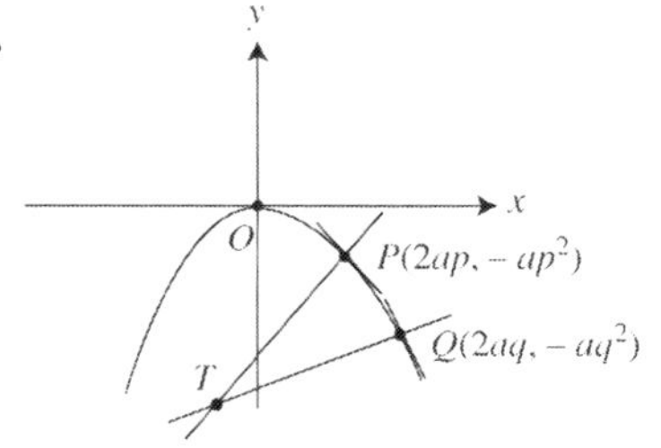

$x^2 = -4ay$ differentiating

with respect to x, $2x = -4a\dfrac{dy}{dx}$

$$\therefore \frac{dy}{dx} = \frac{2x}{-4a} = \frac{2(2ap)}{-4a} = -p$$

the gradient of the of the normal is $\dfrac{1}{p}$

$$\therefore y = \frac{1}{p}x + c \Rightarrow -ap^2 = \frac{1}{p}(2ap) + c \Rightarrow c = -ap^2 - 2a$$

$$\therefore y = \frac{1}{p}x - 2a - ap^2 \quad \ldots(1) \qquad \text{Equation of the normal at } P$$

$$y = \frac{1}{q}x - 2a - aq^2 \quad \ldots(2) \qquad \text{Equation of the normal at } Q$$

Solving (1) and (2) we have

$$\frac{1}{p}x - ap^2 - 2a = \frac{1}{q}x - aq^2 - 2a \Rightarrow x\left(\frac{1}{p} - \frac{1}{q}\right) = ap^2 + \cancel{2a} - aq^2 - \cancel{2a}$$

$$x\frac{(q-p)}{pq} = a(p^2 - q^2) \Rightarrow x = \frac{a(p-q)(p+q)}{(q-p)pq} = -apq(p+q)$$

$$y = \frac{1}{q}\left[-apq(p+q)\right] - aq^2 - 2a = -ap^2 - apq - aq^2 - 2a$$

$$y = -a(p^2 + q^2) - apq - 2a$$

$$\therefore T\left[-apq(p+q), -a(p^2+q^2) - apq - 2a\right].$$

If $pq = -1$, $x = a(p+q)$ and $y = -a(p^2+q^2) + a - 2a$

$$y = -a(p^2+q^2) - a = -a\left[(p+q)^2 - 2pq\right] = -a\left[\left(\frac{x}{a}\right)^2 + 2\right] - a$$

$$y = -a\frac{x^2}{a^2} - 3a = -\frac{ax^2}{a^2} - 3a. \qquad \therefore \text{The locus} \quad \boxed{y = -\frac{x^2}{a} - 3a}$$

---

6. $\mathbf{A} = \begin{pmatrix} \frac{3}{5} & \frac{1}{5} \\ \frac{4}{5} & \frac{3}{5} \end{pmatrix}$ the minors of $\mathbf{A} = \begin{pmatrix} \frac{3}{5} & \frac{4}{5} \\ \frac{1}{5} & \frac{3}{5} \end{pmatrix}$

$$\text{the cofactors of } \mathbf{A} = \begin{pmatrix} \frac{3}{5} & \frac{4}{5} \\ -\frac{1}{5} & \frac{3}{5} \end{pmatrix} = \mathbf{A}^* \qquad \text{the adjoint matrix of } \mathbf{A} = \mathbf{A}^{*T}$$

$$\mathbf{A}^{*T} = \begin{pmatrix} \frac{3}{5} & -\frac{1}{5} \\ -\frac{4}{5} & \frac{3}{5} \end{pmatrix}$$

$$\therefore \mathbf{A}^{-1} = \frac{\mathrm{A}^{*T}}{|\mathbf{A}|} = \frac{\begin{pmatrix} \frac{3}{5} & -\frac{1}{5} \\ -\frac{4}{5} & \frac{3}{5} \end{pmatrix}}{\frac{1}{5}} = 5 = \begin{pmatrix} \frac{3}{5} & -\frac{1}{5} \\ -\frac{4}{5} & \frac{3}{5} \end{pmatrix} = \begin{pmatrix} 3 & -1 \\ -4 & 3 \end{pmatrix}$$

$$|\mathbf{A}| = \frac{3}{5} \times \frac{3}{5} - \frac{4}{5} \times \frac{1}{5} = \frac{1}{5}$$

$$\mathbf{AA}^{-1} = \begin{pmatrix} \frac{3}{5} & \frac{1}{5} \\ \frac{4}{5} & \frac{3}{5} \end{pmatrix} \begin{pmatrix} 3 & -1 \\ -4 & 3 \end{pmatrix}$$

$$= \begin{pmatrix} \frac{9}{5} - \frac{4}{5} & -\frac{3}{5} + \frac{3}{5} \\ \frac{12}{5} - \frac{12}{5} & -\frac{4}{5} + \frac{9}{5} \end{pmatrix} = \begin{pmatrix} 1 & 0 \\ 0 & 1 \end{pmatrix} = \mathbf{I}$$

$$\mathbf{A}^{-1}\mathbf{A} = \begin{pmatrix} 3 & -1 \\ -4 & 3 \end{pmatrix} = \begin{pmatrix} \frac{3}{5} & \frac{1}{5} \\ \frac{4}{5} & \frac{3}{5} \end{pmatrix} = \begin{pmatrix} 1 & 0 \\ 0 & 1 \end{pmatrix} = \mathbf{I}$$

$\therefore$ **A** and $\mathbf{A}^{-1}$ are commutative.

---

7. (i) $z_1 z_2 = (1 + 3i)(\sqrt{3} - i) = \sqrt{3} + 3\sqrt{3}i - i + 3 = 3 + \sqrt{3} + i\,(3\sqrt{3} - 1)$

(ii) $z_1 + z_2 = 1 + 3i + \sqrt{3} - i = 1 + \sqrt{3} + 2i$

(iii) $\dfrac{z_2}{z_1} = \dfrac{\sqrt{3} - i}{1 + 3i} \times \dfrac{1 - 3i}{1 - 3i} = \dfrac{\sqrt{3} - i - 3\sqrt{3}i - 3}{1^2 - (3i)^2}$

$$= \frac{\sqrt{3} - 3 - i\left(3\sqrt{3} + 1\right)}{10} = \frac{\sqrt{3} - 3}{10} - i\frac{\left(3\sqrt{3} + 1\right)}{10}$$

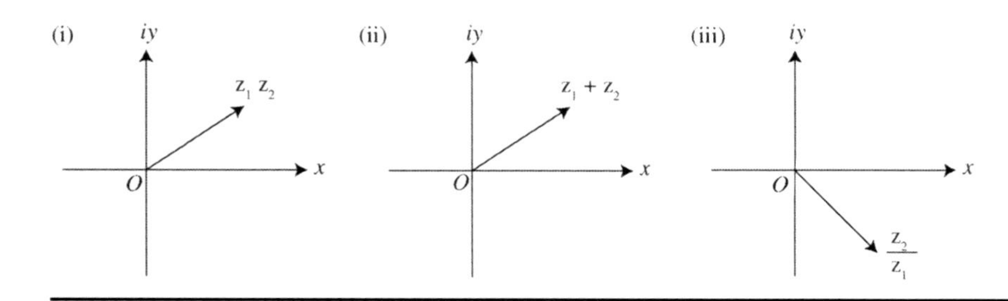

---

8. Let α and β be the roots of the quadratic equation, then $\alpha + \beta = -7$ and $\alpha^3 + \beta^3 = 125$

$$\alpha^3 + \beta^3 = (\alpha + \beta)^3 - 3\alpha^2\beta - 3\alpha\beta^2 = (\alpha + \beta)^3 - 3\alpha\beta(\alpha + \beta)$$

$$125 = (-7)^3 - 3\alpha\beta(-7)$$

$$125 = -343 + 21\alpha\beta$$

$$125 + 343 = 21\alpha\beta$$

$$21\alpha\beta = 468$$

$$\alpha\beta = \frac{156}{7}$$

$$\therefore x^2 - (\alpha + \beta)x + \alpha\beta = 0$$

$$x^2 - (-7)x + \frac{156}{7} = 0$$

$$\boxed{7x^2 + 49x + 156 = 0}$$

---

9. The equation of the straight line AB.

$$y = mx + c \qquad m = \frac{5}{-4} = -\frac{5}{4}$$

$$y = -\frac{5}{4}x + c \quad \text{if } x = 0, \quad y = 5 \Rightarrow c = 5$$

$$y = -\frac{5}{4}x + 5 = 2^x$$

$2^x + \frac{5}{4}x - 5 = 0 \qquad f(x) = 2^x + \frac{5}{4}x - 5$

let $x = 2$, $\qquad f(2) = 4 + 2.5 - 5 = 1.5$

let $x = 3$, $\qquad f(3) = 8 + \frac{15}{4} - 5 = 3 + 3.75 = 6.75$

let $x = 1$, $\qquad f(1) = 2 + 1.25 - 5 = -1.75.$

The interval values in which there is a change of sign is [1, 2]

(i)

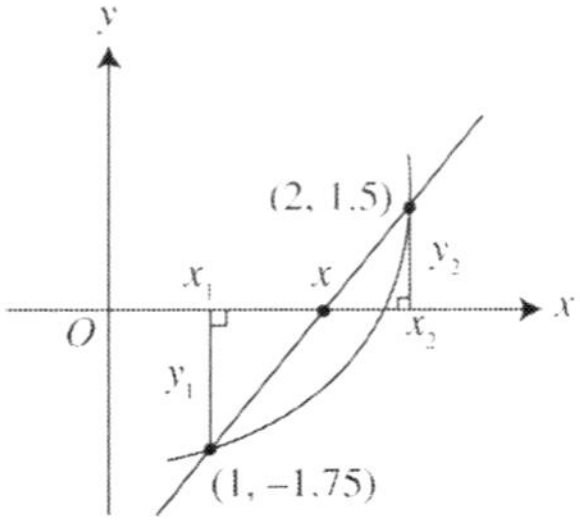

$$\frac{y_2}{x_2 - x} = \frac{y_1}{x - x_1} \Rightarrow \frac{y_2}{y_1} = \frac{x_2 - x}{x - x_1}$$

$$\frac{1.5}{1.75} = \frac{2 - x}{x - 1}$$

$$1.5x - 1.5 = (2 - x)1.75 = 3.5 - 1.75x$$

$$1.5x + 1.75x = 1.5 + 3.5 = 5$$

$$3.25x = 5$$

$$x = \frac{5}{3.25} = 1.538461538$$

$$x = 1.54 \text{ to 3 s.f.}$$

$$f(1.54) = 2^{1.54} + 1.25 \times 1.54 - 5 = -0.167054965$$

(ii) Interval bisection $\qquad \frac{1 + 2}{2} = 1.5$

$$f(1.5) = 2^{1.5} + 1.25(1.5) - 5 = -0.296572875$$

$$\frac{1.5 + 2}{2} = \frac{3.75}{2} = 1.875$$

$$f(1.875) = 2^{1.875} + 1.25 \times 1.875 - 5 = 1.011766173$$

$$\frac{1.875 + 1.5}{2} = \frac{3.375}{2} = 1.6875.$$

(iii) $x_{n+1} = x_n - \dfrac{f(x_n)}{f'(x_n)}$

$$f(x) = 2^x + 1.25x - 5$$

$$f'(x) = 2^x \ln 2 + 1.25$$

$$x_{n+1} = 1.54 - \frac{f(1.54)}{f'(1.54)} = 1.54 - \frac{(-0.167054965)}{3.515633902}$$

$$= 1.547517736 = 1.55 \text{ to 3 s.f.}$$

$f'(1.54) = 2^{1.54} \ln + 1.25 = 3.515633902.$

Examination Test Papers.

GCE Advanced Level.

Further Pure Mathematics FP2

Test Paper 1

Time: 1 hour and 30 minutes.

Instructions and Information

Candidates may use any calculator allowed by the regulations of their Examination Board.

Full marks are awarded for correct answers to ALL questions.

This paper has eight questions.

You can start working with any question and you must label clearly all parts.

1. Plot the following polar coordinates on the same diagram.

$$A\left(3, \frac{\pi}{4}\right), \quad B\left(4, \frac{\pi}{2}\right), \quad C\left(-2, \frac{\pi}{4}\right).$$

Calculate the distances AB, BC, and AC. **(6)**

---

2. Use the expansions of $\sin x$ and $\cos x$ to obtain the expansion of $\tan x$, as far as the term in x^5. Use Maclaurin's Expansion to show that

$\tan x \approx x + \frac{1}{3}x^3 + \cdots$ **(8)**

---

3. Prove $$\sum_{r=1}^{n} \frac{1}{(2r-1)(2r+1)} = \frac{n}{2n+1}$$

and hence evaluate $\displaystyle\sum_{r=15}^{30} \frac{1}{(2r-1)(2r+1)}$ to 3 s.f. **(12)**

---

4. Solve the differential equation

$\dfrac{dy}{dt} = \left(3t - t^{-3}\right)^3 \left(3 + 3t^{-4}\right)$ given that $t = 1$ when $y = 2$. **(8)**

---

5. Prove by the method of mathematical induction that

$(\cos\theta + i\sin\theta)^n = \cos n\theta + i\sin n\theta$ if n is a positive integer. **(6)**

---

6. Draw up a table of values for θ and r at intervals of $\frac{\pi}{6}$, $0 \leq \theta \leq 2\pi$.

Sketch the graph $r = 1 + \cos\theta$.

Determine the area enclosed by the curve and the initial line. **(12)**

---

7. If $Z_1 = \cos\frac{\pi}{4} + i\sin\frac{\pi}{4}$ and $Z_2 = 2\left(\cos\frac{\pi}{3} + i\sin\frac{\pi}{3}\right)$

find (i) Z_1Z_2, (ii) $\frac{Z_1}{Z_2}$ and (iii) $\frac{Z_2}{Z_1}$.

(a) Using the Euler's formula.

(b) Using the Polar form. **(15)**

---

8. Solve the inequalities:

(i) $\dfrac{3x - 2}{x} \geq 3$ (ii) $\dfrac{3}{x - 1} > 1$ **(8)**

TOTAL FOR PAPER: 75 MARKS

Examination Test Papers.

GCE Advanced Level.

Test Paper 1 Solutions

Further Pure Mathematics FP2

1.

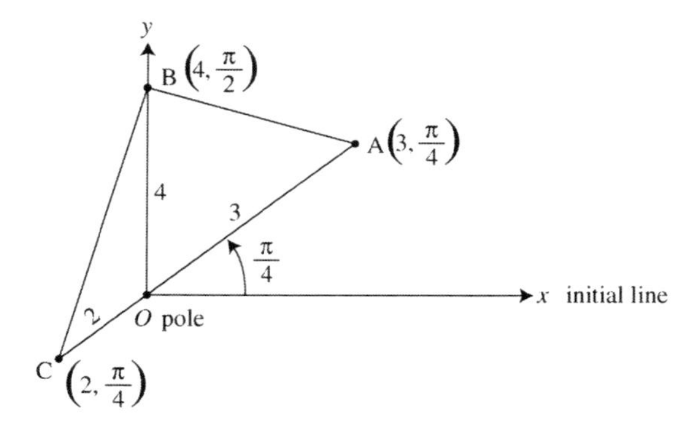

Using the cosine rule for the ΔOAB

$$AB^2 = OA^2 + OB^2 - 2 \times (OA) \times (OB) \cos \widehat{AOB}$$

$$= 3^2 + 4^2 - 2 \times 3 \times 4 \times \cos 45^\circ = 9 + 16 - 24 \times \frac{1}{\sqrt{2}} = 25 - 16.97056275$$

$AB = \sqrt{8.029437252}$ $\quad AB = \sqrt{8.03}$ to 3 s.f.

AB = 2.83 units to 3 s.f.

ΔOBC

$$BC^2 = OB^2 + OC^2 - 2 \times 2 \times 4 \times \cos 135^\circ = 16 + 4 - 16 \times (-0.707) = 31.3137085$$

$$BC = \sqrt{31.3137085} = 5.595865304$$

$BC = 5.60$ units

$AC = 2 + 3 = 5.$

---

2. $\tan x = \dfrac{\sin x}{\cos x} = \dfrac{x - \frac{x^3}{3!} + \frac{x^5}{5!}}{1 - \frac{x^2}{2!} + \frac{x^4}{4!}}$ using long division

$$x + \frac{x^3}{3} + \frac{2}{15}x^5$$

$$1 - \frac{x^2}{2!} + \frac{x^4}{4!} \,\Big)\, x - \frac{x^3}{3!} + \frac{x^5}{5!}$$

$$x - \frac{x^3}{2!} + \frac{x^5}{4!}$$

$$\frac{1}{3}x^3 - \frac{1}{30}x^5$$

$$\frac{1}{3}x^3 - \frac{x^5}{6}$$

$$\frac{2}{15}x^5$$

$\therefore \tan x = x + \frac{x^3}{3} + \frac{2}{15}x^5$

$$f(x) = f(0) + f'(0)\frac{x}{1!} + f''(0)\frac{x^2}{2!} + f'''(0)\frac{x^3}{3!} + \dots$$

$$f(x) = \tan x \qquad f(0) = 0$$

$$f'(x) = \sec^2 x \quad f'(0) = 1$$

$$f''(x) = 2\sec x \sec x \tan x = 2\sec^2 x \tan x \qquad f''(0) = 0$$

$$f'''(x) = 4\sec x \sec x \tan^2 x + 2\sec^2 x \sec^2 x \quad f'''(0) = 2$$

$$\therefore \tan x = x + \frac{x^3}{3} + \cdots$$

---

3. Let

$$\frac{1}{(2r-1)\ (2r+1)} \equiv \frac{A}{2r-1} + \frac{B}{2r+1}$$

$$1 \equiv A(2r+1) + B(2r-1).$$

$$\text{If } r = -\frac{1}{2},\quad B = -\frac{1}{2}.$$

$$\text{If } r = \frac{1}{2},\quad A = \frac{1}{2}.$$

$$\therefore \frac{1}{(2r-1)\ (2r+1)} \equiv \frac{1}{2(2r-1)} - \frac{1}{2(2r+1)}$$

$$\sum_{r=1}^{n} \frac{1}{(2r-1)\ (2r+1)} = \frac{1}{2}\sum_{r=1}^{n}\frac{1}{2r-1} - \frac{1}{2}\sum_{r=1}^{n}\frac{1}{2r+1}$$

$$= \frac{1}{2}\left(\frac{1}{1} + \frac{1}{3} + \frac{1}{5} + \cdots + \frac{1}{2n-1}\right)$$

$$-\frac{1}{2}\left(\frac{1}{3} + \frac{1}{5} + \frac{1}{7} + \cdots + \frac{1}{2n+1}\right)$$

Let $\frac{1}{3} + \frac{1}{5} + \frac{1}{7} + \cdots + \frac{1}{2n-1} = S$, and the term before the $\frac{1}{2n+1}$ is $\frac{1}{2(n-1)+1} = \frac{1}{2n-1}$

$$\therefore \sum_{r=1}^{n} \frac{1}{(2r-1)(2r+1)} = \frac{1}{2}\left(\frac{1}{1} + S\right) - \frac{1}{2}\left(S + \frac{1}{2n+1}\right)$$

$$= \frac{1}{2} - \left(\frac{1}{2(2n+1)}\right) = \frac{2n+1-1}{2(2n+1)} = \frac{2n}{2(2n+1)} = \frac{n}{2n+1}.$$

$$\sum_{r=1}^{30} \frac{1}{(2r-1)(2r+1)} = \sum_{r=1}^{14} \frac{1}{(2r-1)(2r+1)} + \sum_{r=15}^{30} \frac{1}{(2r-1)(2r+1)}.$$

$$\sum_{r=15}^{30} \frac{1}{(2r-1)(2r+1)} = \sum_{r=1}^{30} \frac{1}{(2r-1)(2r+1)} - \sum_{r=1}^{14} \frac{1}{(2r-1)(2r+1)}.$$

$$= \frac{30}{61} - \frac{14}{29} = 9.04 \times 10^{-3} \text{ to 3 s.f.}$$

---

4. $\frac{dy}{dt} = \left(3t - t^{-3}\right)^3 \left(3 + 3t^{-4}\right)$

$dy = \left(3t - t^{-3}\right)^3 \left(3 + 3t^{-4}\right) \text{ dt} \qquad \ldots(1)$

let $u = 3t - t^{-3}$ $\quad \frac{du}{dt} = 3 + 3t^{-4}$

integrating both sides of (1)

$$y = \int \left(3t - t^{-3}\right)^3 \left(3 + 3t^{-4}\right) dt$$

$$= \int u \frac{du}{3 + 3t^{-4}} \left(3 + 3t^{-4}\right)$$

$$= \int u \, du = \frac{u^2}{2} + c = \frac{1}{2}\left(3t - t^{-3}\right)^2 + c$$

$$2 = \frac{1}{2}\left(3 \times 1 - 1^{-3}\right)^2 + c \Rightarrow c = 0$$

$$\therefore y = \frac{1}{2}\left(3t - t^{-3}\right)^2.$$

---

5. $(\cos\theta + i\sin\theta)^n = (\cos n\theta + i\sin n\theta)$.

For $n = 1$ $\quad (\cos\theta + i\sin\theta)^1 = \cos\theta + i\sin\theta$ $\quad$ it is true

For $n = k$ $\quad (\cos\theta + i\sin\theta)^k = \cos k\theta + i\sin k\theta$ $\quad \ldots (1)$ $\quad$ it would be true.

For $n = k+1$ $\quad (\cos\theta + i\sin\theta)^{k+1} = (\cos\theta + i\sin\theta)^k\,(\cos\theta + i\sin\theta)$

$$= (\cos k\theta + i\sin k\theta)\,(\cos\theta + i\sin\theta)$$

$$= \cos k\theta\cos\theta + i\sin k\theta\cos\theta + i\sin\theta\cos k\theta - \sin k\theta\sin\theta$$

$$= (\cos k\theta\cos\theta - \sin k\theta\sin\theta) + i\,(\sin k\theta\cos\theta - \sin k\theta\sin\theta)$$

$$= \cos(k+1)\,\theta + i\sin(k+1)\,\theta$$

$\therefore$ it is true for n.

6.

θ^c	0	$\frac{\pi}{6}$	$\frac{\pi}{3}$	$\frac{\pi}{2}$	$\frac{2\pi}{3}$,	$\frac{5\pi}{6}$	π	$\frac{7\pi}{3}$	$\frac{4\pi}{3}$	$\frac{3\pi}{2}$	$\frac{5\pi}{3}$	$\frac{11\pi}{6}$	2π
$\cos\theta$	1	$\frac{\sqrt{3}}{2}$	$\frac{1}{2}$	0	$-\frac{1}{2}$	$-\frac{\sqrt{3}}{2}$	-1	$-\frac{\sqrt{3}}{2}$	$-\frac{1}{2}$	0	$\frac{1}{2}$	$\frac{\sqrt{3}}{2}$	1
$r = 1 + \cos\theta$	2	$1 + \frac{\sqrt{3}}{2}$	$\frac{3}{2}$	1	$\frac{1}{2}$	$1 - \frac{\sqrt{3}}{2}$	0	$1 - \frac{\sqrt{3}}{2}$	$\frac{1}{2}$	1	$\frac{3}{2}$	$1 + \frac{3}{2}$	2

$$\frac{1}{2}\int_0^{\pi} r^2\,d\theta$$

$$= \frac{1}{2}\int_0^{\pi} (1 + \cos\theta)^2\,d\theta$$

$$= \frac{1}{2}\int_0^{\pi} (1 + 2\cos\theta + \cos^2\theta)\,d\theta$$

$$\cos 2\theta = 2\cos^2\theta - 1 \Rightarrow \frac{\cos 2\theta + 1}{2} = \cos^2\theta.$$

Area hatched

$$= \frac{1}{2}\int_0^{\pi}\left[1 + 2\cos\theta + \frac{\cos 2\theta + 1}{2}\right]d\theta = \frac{1}{2}\left[\theta + 2\sin\theta + \frac{\sin 2\theta}{4} + \frac{1}{2}\theta\right]_0^{\pi} = \frac{1}{2}\frac{3}{2}\pi$$

$$= \frac{3}{4}\pi \text{ s.u.}$$

---

7. (a) Euler's formula

$$e^{i\theta} = \cos\theta + i\sin\theta$$

$$Z_1 = \cos\frac{\pi}{4} + i\sin\frac{\pi}{4} = 1e^{i\frac{\pi}{4}} = e^{i\frac{\pi}{4}}$$

$$Z_2 = 2\left(\cos\frac{\pi}{3} + i\sin\frac{\pi}{3}\right) = 2e^{i\frac{\pi}{3}}$$

(i) $Z_1 Z_2 = e^{i\frac{\pi}{4}}\ 2e^{i\frac{\pi}{3}} = 2e^{i\left(\frac{\pi}{4}+\frac{\pi}{3}\right)} = 2e^{i\frac{7\pi}{12}}$

(ii) $\frac{Z_1}{Z_2} = \frac{e^{i\frac{\pi}{4}}}{2e^{i\frac{\pi}{3}}} = \frac{1}{2}e^{i\frac{\pi}{4}}e^{-i\frac{\pi}{3}} = \frac{1}{2}e^{-i\frac{\pi}{12}}$

(iii) $\frac{Z_2}{Z_1} = \frac{2e^{i\frac{\pi}{3}}}{e^{i\frac{\pi}{4}}} = 2e^{i\frac{\pi}{12}}$.

(b) Polar form.

$Z = \cos\theta + i\sin\theta \quad |Z| = 1 \quad \arg Z = \theta$

$Z = 1(\cos\theta + i\sin\theta) = 1\angle\theta$

where $\angle\theta = \cos\theta + i\sin\theta$

$Z_1 = \cos\frac{\pi}{4} + i\sin\frac{\pi}{4}. = \frac{1}{\sqrt{2}} + i\frac{1}{\sqrt{2}}$

$|Z_1| = \sqrt{\left(\frac{1}{\sqrt{2}}\right)^2 + \left(\frac{1}{\sqrt{2}}\right)^2} = 1$

$\arg Z_1 = \tan^{-1}\frac{\frac{1}{\sqrt{2}}}{\frac{1}{\sqrt{2}}} = \tan^{-1}1 = \frac{\pi}{4}$

$Z_1 = 1\angle\frac{\pi}{4}$ and $Z_2 = 2\angle\frac{\pi}{3} = 2\left(\cos\frac{\pi}{3} + i\sin\frac{\pi}{3}\right) = 2\left(\frac{1}{2} + i\frac{\sqrt{3}}{2}\right) = 1 + i\sqrt{3}$

$|Z_2| = \left|1 + i\sqrt{3}\right| = \sqrt{(1)^2 + \left(\sqrt{3}\right)^2} = 2$

$\arg Z_2 = \frac{\pi}{3}$

(i) $Z_1 Z_2 = |Z_1||Z_2|\angle\frac{\pi}{4} + \frac{\pi}{3}$

$= 1 \times 2\angle\frac{7\pi}{12} = 2\left(\cos\frac{7\pi}{12} + i\sin\frac{7\pi}{12}\right).$

(ii) $\dfrac{z_1}{z_2} = \dfrac{1\angle\frac{\pi}{4}}{2\angle\frac{\pi}{3}} = \dfrac{1}{2}\angle\frac{\pi}{4} - \frac{\pi}{3}$

$= \dfrac{1}{2}\angle -\frac{\pi}{12} = \dfrac{1}{2}\left(\cos\frac{\pi}{12} - i\sin\frac{\pi}{12}\right)$

(iii) $\dfrac{z_2}{z_1} = \dfrac{2\angle\frac{\pi}{3}}{1\angle\frac{\pi}{4}} = 2\angle\frac{\pi}{3} - \frac{\pi}{4}$

$= 2\left(\cos\frac{\pi}{12} + i\sin\frac{\pi}{12}\right).$

---

8. (i) $\dfrac{3x-2}{x} \geq 3$ multiplying both sides by x^2

$\dfrac{3x-2}{x}x^2 \geq 3x^2$

$x(3x-2) \geq 3x^2$

$3x^2 - 2x - 3x^2 \geq 0$

$-2x \geq 0 \quad 2x \leq 0 \quad \boxed{x \leq 0}$

(ii) $\dfrac{3}{x-1} > 1$ multiplying both sides by $(x-1)^2$

$\dfrac{3(x-1)^2}{x-1} > (x-1)^2 \quad 3(x-1) - (x-1)^2 > 0 \quad (x-1)[3-(x-1)] > 0$

$(x-1)(4-x) > 0$ or $(x-1)(x-4) < 0$

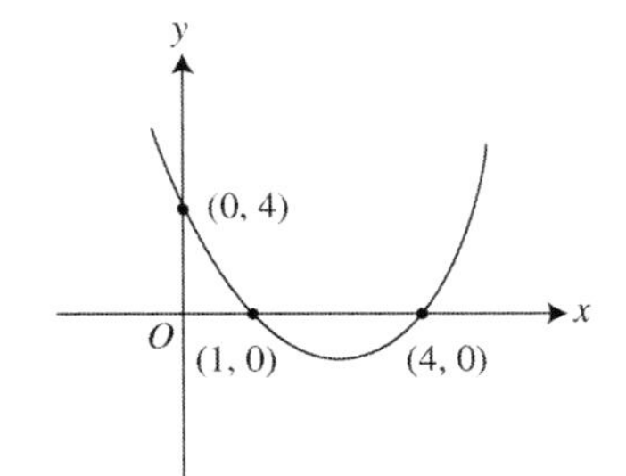

$\boxed{1 < x < 4}$

Examination Test Papers.

GCE Advanced Level.

Further Pure Mathematics FP2

Test Paper 2

Time: 1 hour and 30 minutes.

Instructions and Information

Candidates may use any calculator allowed by the regulations of their Examination Board.

Full marks are awarded for correct answers to ALL questions.

This paper has eight questions.

You can start working with any question and you must label clearly all parts.

1. The diagonals of a rhombus $ABCD$ intersect at the pole, the polar coordinates of A and B are $(1, 0^c)$ and $\left(4, \frac{\pi}{2}\right)$ respectively, determine the following:-

(i) The polar coordinates of the other two points C and D.

(ii) The distance AB.

(iii) The area of the rhombus.

(iv) If the rhombus is rotated clockwise by $\frac{\pi}{4}$, about the pole, find the polar coordinates of A, B, C and D, in the range $-\pi \leq 0 \leq \pi$. **(12)**

---

2. Use Maclaurin's series to expand $e^{\sin x}$ as far as the term in x^4. **(8)**

---

3. a) Express $\sum_{r=1}^{n} \frac{1}{r(r+1)(r+2)}$ into partial fractions.

b) Prove that $\sum_{r=1}^{n} \frac{1}{r(r+1)(r+2)} = \frac{1}{4} - \frac{1}{2(n+1)} + \frac{1}{2(n+2)}$.

c) Evaluate $\sum_{r=10}^{20} = \frac{1}{r(r+1)(r+2)}$ to 3 s.f. **(12)**

---

4. Find the general solution of the exact differential equation

$e^y \tan x \frac{dy}{dx} + \sec^2 x \, e^y = x \ln x$. **(6)**

---

5. (a) Use the binomial theorem to expand $(\cos\theta + i \sin\theta)^n$, let $\cos\theta = c$ and $\sin\theta = s$, hence use Demoivre's theorem to find expressions for $\cos n\theta$ and $\sin n\theta$.

(b) Simplify $\dfrac{(\cos\theta + i \sin\theta)^3}{(\sin\theta + i \cos\theta)^4}$ giving your answers in terms of a multiple angle of θ. **(9)**

---

6. For the table of values shown

θ^c: $0 \quad \frac{\pi}{6} \quad \frac{\pi}{3} \quad \frac{\pi}{2} \quad \frac{2\pi}{3} \quad \frac{5\pi}{6} \quad \pi \quad \frac{7\pi}{6} \quad \frac{4\pi}{3} \quad \frac{3\pi}{2} \quad \frac{5\pi}{3} \quad \frac{11\pi}{6} \quad 2\pi$.

Sketch $r = \sin 3\theta$.

Determine the area of one lobe. **(8)**

---

7. If $Z_1 = -2\sqrt{3} + i2$ and $Z_2 = \sqrt{3} + i\sqrt{2}$.

Express in polar form:

(a) Z_1 and Z_2 (b) Z_1Z_2 (c) $\dfrac{Z_1}{Z_2}$.

Giving your answers in degrees and in 3 s.f. **(14)**

---

8. Find the set of real values of x for which $\dfrac{x-1}{x(x+1)} < 0$. **(6)**

TOTAL FOR PAPER: 75 MARKS

Examination Test Papers.

GCE Advanced Level.

Test Paper 2 Solutions

Further Pure Mathematics FP2

1. (i)

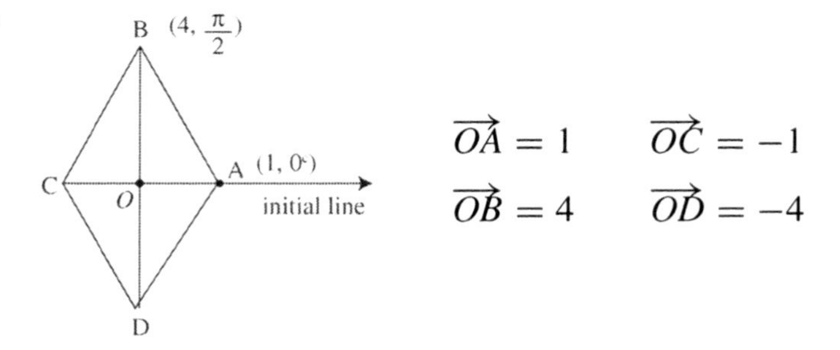

$$\overrightarrow{OA} = 1 \qquad \overrightarrow{OC} = -1$$
$$\overrightarrow{OB} = 4 \qquad \overrightarrow{OD} = -4$$

The diagonals of the rhombus intersect at 90°

$$C\,(-1, 0^c) \quad \text{or} \quad C\,(1, \pi^c)$$
$$D\left(-4, \tfrac{\pi}{2}\right) \quad \text{or} \quad D\left(4, \tfrac{3\pi}{2}\right) \quad \text{or} \quad D\left(4, -\tfrac{\pi}{2}\right)$$

(ii) $AB^2 = OB^2 + OA^2 = 1 + 16 = 17$

$AB = \sqrt{17}$ units.

(iii) Area of rhombus = $4 \times \frac{1}{2}\,(1 \times 4) = 8$ s.u.

(iv) The new values of the coordinates are:

$$A\left(1, -\frac{\pi}{4}\right), \quad B\left(4, \frac{\pi}{4}\right), \quad C\left(1, \frac{3\pi}{4}\right) \quad \text{and} \quad D\left(4, -\frac{3\pi}{4}\right).$$

---

2. $e^{\sin x} = e^{(x - \frac{x^3}{3!} + \frac{x^5}{5!} - \frac{x^7}{7!} \cdots)}$

$$= e^x e^{-\frac{x^3}{3!}} e^{+\frac{x^5}{5!}} e^{-\frac{x^7}{7!}}$$

$$= \left(1 + x + \frac{x^2}{2!} + \frac{x^3}{3!} + \frac{x^4}{4!}\right)\left(1 - \frac{x^3}{3!}\right)\left(1 + \frac{x^5}{5!}\right)\left(1 - \frac{x^7}{7!}\right)$$

$$= \left(1 + x + \frac{x^2}{2!} + \frac{x^3}{3!} + \frac{x^4}{4!} - \frac{x^3}{3!} - \frac{x^4}{3!}\right)\left(1 + \frac{x^5}{5!}\right)\left(1 - \frac{x^7}{7!}\right)$$

$$= 1 + x + \frac{1}{2}x^2 + \left(\frac{1}{4!} - \frac{1}{3!}\right)x^4$$

$$= 1 + x + \frac{1}{2}x^2 - \frac{1}{8}x^4.$$

---

3. (a) $\frac{1}{r(r+1)(r+2)} \equiv \frac{A}{r} + \frac{B}{r+1} + \frac{C}{r+2}$

$$1 \equiv A(r+1)(r+2) + Br(r+2) + Cr(r+1).$$

If $r = -1$, $B = -1$, if $r = 0$, $A = \frac{1}{2}$ and if $r = -2$, $C = \frac{1}{2}$.

$$\therefore \frac{1}{r(r+1)(r+2)} = \frac{1}{2r} - \frac{1}{r+1} + \frac{1}{2(r+2)}.$$

(b) $\sum_{r=1}^{n} \frac{1}{r(r+1)(r+2)} = \frac{1}{2\times 1} + \frac{1}{2\times 2} + \dots + \frac{1}{2\times n} - \left(\frac{1}{2} + \frac{1}{3} + \frac{1}{4} + \dots + \frac{1}{n+1}\right)$

$$+\frac{1}{2}\left(\frac{1}{3} + \frac{1}{4} + \frac{1}{5} + \dots + \frac{1}{n+2}\right)$$

$$= \frac{1}{2}\left(\frac{1}{1} + \frac{1}{2} + \frac{1}{3} + \dots + \frac{1}{n}\right) - \left(\frac{1}{2} + \frac{1}{3} + \frac{1}{4} + \dots + \frac{1}{n+1}\right)$$

$$+\frac{1}{2}\left(\frac{1}{3} + \frac{1}{4} + \dots + \frac{1}{n+1} + \frac{1}{n+2}\right)$$

let $\frac{1}{3} + \frac{1}{4} + \frac{1}{5} + \dots + \frac{1}{n} = S$

$$\sum_{r=1}^{n} \frac{1}{r(r+1)(r+2)} = \frac{1}{2}\left(\frac{1}{1} + \frac{1}{2} + S\right) - \left(\frac{1}{2} + S + \frac{1}{n+1}\right) + \frac{1}{2}\left(S + \frac{1}{n+1} + \frac{1}{n+2}\right)$$

$$= \frac{3}{4} + \frac{1}{2}S - \frac{1}{2} - S - \frac{1}{n+1} + \frac{1}{2}S + \frac{1}{2(n+1)} + \frac{1}{2(n+2)}$$

$$= \frac{3}{4} - \frac{1}{2} - \frac{1}{2(n+1)} + \frac{1}{2(n+2)} = \frac{1}{4} - \frac{1}{2(n+1)} + \frac{1}{2(n+2)}.$$

(c) $\sum_{r=1}^{20} \frac{1}{r(r+1)(r+2)} = \sum_{r=1}^{9} \frac{1}{r(r+1)(r+2)} + \sum_{r=10}^{20} \frac{1}{r(r+1)(r+2)}$

$$\therefore \sum_{r=10}^{20} \frac{1}{r(r+1)(r+2)} = \sum_{r=1}^{20} \frac{1}{r(r+1)(r+2)} - \sum_{r=1}^{9} \frac{1}{r(r+1)(r+2)}$$

$$= \frac{1}{4} - \frac{1}{2\times 21} + \frac{1}{2\times 22} - \frac{1}{4} + \frac{1}{2\times 10} - \frac{1}{2\times 11}$$

$$= -\frac{1}{42} + \frac{1}{44} + \frac{1}{20} - \frac{1}{22} = 3.46 \times 10^{-3} \text{ to 3 s.f.}$$

---

4. $e^y \tan x \frac{dy}{dx} + \sec^2 x \, e^y = x \ln x$

$$\frac{d}{dx}(e^y \tan x) = e^y \frac{dy}{dx} \tan x + e^y \sec^2 x$$

$$\frac{d}{dx}(e^y \tan x) = x \ln x$$

$$d(e^y \tan x) = x \ln x \, dx \cdots (2)$$

$$\int d\left(e^y \tan x\right) = \int \underset{①}{x} \; \underset{②}{\ln x} \, dx$$

$$e^y \tan x = \frac{x^2}{2} \ln x - \int \frac{x^2}{2} \cdot \frac{1}{x} \, dx$$

$$e^y \tan x = \frac{x^2}{2} \ln x - \frac{x^2}{4} + c.$$

---

5. (a) $(\cos\theta + i \sin\theta)^n = (c + is)^n$

$$= c^n + nc^{n-1} is + \frac{n(n-1)}{2!} c^{n-2} i^2 s^2$$

$$+ \frac{n(n-1)(n-2)}{3!} c^{n-3} i^3 s^3 + \cdots + i^n s^n$$

$$= \left(c^n - \frac{n(n-1)}{2!} c^{n-2} s^2 + \cdots\right) + i\left(nc^{n-1} s - \frac{n(n-1)(n-2)}{3!} s^3 \cdots\right)$$

$$= \cos n\theta + i \sin n\theta$$

$$\therefore \cos n\theta = c^n - \frac{n(n-1)}{2!} c^{n-2} s^2 + \frac{n(n-1)(n-2)(n-3)}{4!} c^{n-4} s^4 - \cdots \text{ real terms}$$

$$\sin n\theta = nc^{n-1} s - \frac{n(n-1)(n-2)}{3!} s^3 + \cdots \text{ imaginary terms}$$

(b) $$\frac{(\cos\theta + i \sin\theta)^3}{(\sin\theta + i \cos\theta)^4} = \frac{(\cos\theta + i \sin\theta)^3}{[i(\cos\theta - i \sin\theta)]^4} = \frac{(\cos\theta + i \sin\theta)^3}{i^4 (\cos\theta - i \sin\theta)^4}$$

$$= \frac{(\cos\theta + i \sin\theta)^3}{(\cos\theta + i \sin\theta)^{-4}}$$

$$= (\cos\theta + i \sin\theta)^7 = \cos 7\theta + i \sin 7\theta.$$

6.

θ^c	0	$\frac{\pi}{6}$	$\frac{\pi}{3}$	$\frac{\pi}{2}$	$\frac{2\pi}{3}$	$\frac{5\pi}{6}$	π	$\frac{7\pi}{6}$	$\frac{4\pi}{3}$	$\frac{3\pi}{2}$	$\frac{5\pi}{3}$	$\frac{11\pi}{6}$	2π
$r = \sin 3\theta$	0	1	0	−1	0	1	0	−1	0	1	0	−1	0

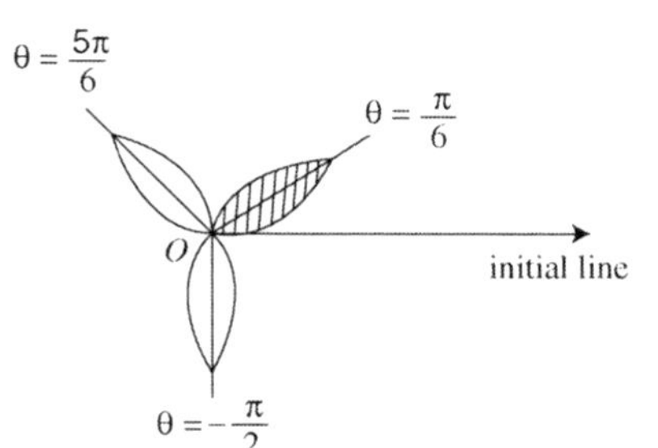

Area of the shaded lobe

$$= \frac{1}{2}\int_0^{\frac{\pi}{3}} r^2 \mathrm{d}\theta$$

$$= \frac{1}{2}\int_0^{\frac{\pi}{3}} \sin^2 3\theta \ \mathrm{d}\theta = \frac{1}{2}\int_0^{\frac{\pi}{3}} \frac{1 - \cos 6\theta}{2}\mathrm{d}\theta$$

where

$$\cos 6\theta = 1 - 2\sin^2 3\theta$$

$$\sin^2 3\theta = \frac{1 - \cos 6\theta}{2}.$$

Area required $= \frac{1}{4}\int_0^{\frac{\pi}{3}} (1 - \cos 6\theta)\ \mathrm{d}\theta = \frac{1}{4}\left[\theta - \frac{\sin 6\theta}{6}\right]_0^{\frac{\pi}{3}} = \frac{1}{4}\left[\frac{\pi}{3}\right] = \frac{\pi}{12}$ s.u.

---

7. (a) $Z_1 = -2\sqrt{3} + i2$

$$|Z_1| = \sqrt{\left(-2\sqrt{3}\right)^2 + 2^2} = \sqrt{12 + 4} = \sqrt{16} = 4$$

$$\arg Z_1 = 180° - \tan^{-1}\frac{2}{2\sqrt{3}}$$

$$= 180° - \tan^{-1}\frac{1}{\sqrt{3}}$$

$$= 180° - 30°$$

$$= 150°$$

$$\therefore |Z_1| = 4\angle 150° = 4\,(\cos 150° + i \sin 150°)$$

$$= 4\cos 150° + i4 \sin 150°$$

$$Z_2 = \sqrt{3} + i\sqrt{2} \qquad |Z_2| = \sqrt{\left(\sqrt{3}\right)^2 + \left(\sqrt{2}\right)^2} = \sqrt{5}$$

$$\arg Z_2 = \tan^{-1}\frac{\sqrt{2}}{\sqrt{3}} = \theta$$

$$Z_2 = \sqrt{5}\ \underline{/\tan^{-1}\tfrac{\sqrt{2}}{\sqrt{3}}} = \sqrt{5}\,(\cos\theta + i\sin\theta)$$

$$= \sqrt{5}\left(\cos\tan^{-1}\frac{\sqrt{2}}{\sqrt{3}} + i\sin\tan^{-1}\frac{\sqrt{2}}{\sqrt{3}}\right)$$

$$= \sqrt{5}\,(\cos 39.2^\circ + i\sin 39.2^\circ) = \sqrt{5}\ \underline{/39.2^\circ}$$

(b) $Z_1Z_2 = 4\,\underline{/150^\circ}\ \ \sqrt{5}\ \underline{/39.2^\circ}$

$$= 4\sqrt{5}\ \underline{/189.2^\circ}$$

$$= 4\sqrt{5}\,(\cos 189.2^\circ + i\sin 189.2^\circ)$$

$$= 8.94\ \underline{/189^\circ}$$

(c) $\dfrac{Z_1}{Z_2} = \dfrac{4\,\underline{/150^\circ}}{\sqrt{5}\,\underline{/39.2^\circ}} = 1.79\ \underline{/110.8^\circ}$

$$= 1.79\,(\cos 110.8^\circ + i\sin 110.8^\circ) = 1.79\ \underline{/111^\circ}.$$

---

8. $\dfrac{x-1}{x(x+1)} < 0$ multiplying both sides by $x^2(x+1)^2$

$$\frac{(x-1)\,x^2\,(x+1)^2}{x(x+1)} < 0 \Rightarrow (x-1)\,x\,(x+1) < 0$$

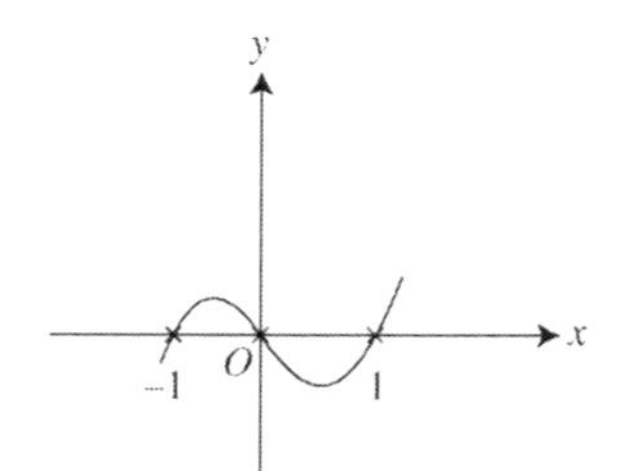

$$f(x) = x(x-1)(x+1)$$

$$f\left(\frac{1}{2}\right) = \frac{1}{2}\left(-\frac{1}{2}\right)\frac{3}{2} < 0$$

$\therefore 0 < x < 1$ and $x < -1$.

Examination Test Papers.

GCE Advanced Level.

Further Pure Mathematics FP2

Test Paper 3

Time: 1 hour and 30 minutes.

Instructions and Information

Candidates may use any calculator allowed by the regulations of their Examination Board.

Full marks are awarded for correct answers to ALL questions.

This paper has eight questions.

You can start working with any question and you must label clearly all parts.

1. Find the distance between the points $A\left(1, \frac{\pi}{8}\right)$ and $B\left(2, \frac{3\pi}{4}\right)$.

Determine the area of the triangle OAB.

---

2. (a) Express $\ln(1+x)$ as power series using Maclaurin's theorem as fa

the term in x^4.

(b) If $\sin x = x - \frac{x^3}{3!} + \frac{x^5}{5!} - \frac{x^7}{7!} + \cdots$ Find the expansion of $\cos x$.

---

3. $\frac{1}{(r+2)(r+4)} \equiv \frac{1}{2(r+2)} - \frac{1}{2(r+4)}$

show that $\sum_{r=1}^{n} \frac{1}{(r+2)(r+4)} = \frac{7}{24} - \frac{1}{2}\frac{2n+7}{(n+3)(n+4)}$

and hence show that $\sum_{r=1}^{50} \frac{1}{(r+2)(r+4)} \approx 0.273$ to 3 s.f.

---

4. Solve the differential equation $x\frac{dy}{dx} + y = x\,e^{-x}$ if $x = 0$ then y

---

5. Using Demoivre's theorem prove that $\tan 3\theta = \frac{3\tan\theta - \tan^3\theta}{1 - 3\tan^2\theta}$.

---

6. Prove the binomial expansion

$$(a+b)^n = a^n + na^{n-1}b + \frac{n(n-1)}{2!}a^{n-2}b^2 + \frac{n(n-1)(n-2)}{3!} \times a^{n-3}b^3 + \cdots$$

using Taylor's theorem. **(9)**

---

7. If $z_1 = 3 + i4$ and $z_2 = 5 + i12$ find (i) $z_1 z_2$ (ii) $\frac{z_1}{z_2}$.

using the polar form.

Expressing your results in radians and in 3 s.f. **(8)**

---

8. Solve the inequalities:

(i) $(x-1)(x+2)(x-3) < 0$

(ii) $x(x+1)(x-2) > 0$. **(12)**

TOTAL FOR PAPER: 75 MARKS

Examination Test Papers.

GCE Advanced Level.

Test Paper 3 Solutions

Further Pure Mathematics FP2

1. The points A and B are plotted as shown in the diagram

$$\widehat{AOB} = \frac{3\pi}{4} - \frac{\pi}{8} = \frac{5\pi}{8}$$

$$AB^2 = OA^2 + OB^2 - 2\,(OA)\,(OB)\cos\frac{5\pi}{8} \quad \text{using the cosine rule}$$

$$= 1^2 + 2^2 - 2\,(1)\,(2)\cos\frac{5\pi}{8} = 1^2 + 2^2 - 2\,(1)\,(2)\,(-0.383\ldots)$$

$$= 6.530733729 \qquad AB = 2.56 \text{ units to 3 s.f.}$$

The area of the triangle $= \dfrac{1}{2}\,(OA)(OB)\sin\dfrac{5\pi}{8} = \dfrac{1}{2}\,(1)\,(2) \times 0.923879532$

$= 0.924$ s.u. to 3 s.f.

---

2. (a)

$$f(x) = \ln(1+x) \qquad f(0) = 0$$

$$f'(x) = \frac{1}{1+x} \qquad f'(0) = 1$$

$$f''(x) = -\frac{1}{(1+x)^2} \qquad f''(0) = -1$$

$$f'''(x) = \frac{2}{(1+x)^3} \qquad f'''(0) = 2$$

$$f^{(iv)}(x) = -\frac{6}{(1+x)^4} \qquad f^{(iv)}(0) = -6$$

$$\therefore \ln(1+x) = x - \frac{x^2}{2} + \frac{x^3}{3} - \frac{x^4}{4}.$$

(b) $\sin x = x - \dfrac{x^3}{3!} + \dfrac{x^5}{5!} - \cdots$

$$\frac{\mathrm{d}}{\mathrm{dx}}(\sin x) = \cos x = 1 - \frac{3x^2}{3!} + \frac{5x^4}{5!} - \cdots$$

$$\therefore \cos x = 1 - \frac{x^2}{2!} + \frac{x^4}{4!} - \cdots$$

---

3. $$\sum_{r=1}^{n}\frac{1}{(r+2)(r+4)}=\frac{1}{2}\left(\frac{1}{3}+\frac{1}{4}+\frac{1}{5}+...+\frac{1}{n+2}\right)-\frac{1}{2}\left(\frac{1}{5}+\frac{1}{6}+....+\frac{1}{n+4}\right)$$

let $S=\frac{1}{5}+\frac{1}{6}+...+\frac{1}{n+2}$

$$=\frac{1}{2}\left(\frac{1}{3}+\frac{1}{4}+S\right)-\frac{1}{2}\left(S+\frac{1}{n+3}+\frac{1}{n+4}\right)$$

$$=\frac{1}{2}\left(\frac{1}{3}+\frac{1}{4}\right)+\frac{1}{2}S-\frac{1}{2}S-\frac{1}{2(n+3)}-\frac{1}{2(n+4)}$$

$$=\frac{7}{24}-\frac{1}{2(n+3)}-\frac{1}{2(n+4)}$$

$$=\frac{7}{24}-\frac{1}{2}\frac{2n+7}{2(n+3)(n+4)}$$

$$\sum_{r=1}^{50}=\frac{7}{24}-\frac{1}{2}\times\frac{107}{53\times 54}=0.272973445=0.273 \text{ to 3 s.f.}$$

---

4. $x\frac{dy}{dx}+y=xe^{-x}$.

This can be written in the form $\frac{dy}{dx}+Py=Q$ where P and Q are functions of x.

$$\frac{dy}{dx}+\frac{1}{x}y=e^{-x} \qquad ...(1)$$

where $P=\frac{1}{x}$ and $Q=e^{-x}$.

The I.F. $=e^{\int p\,dx}=e^{\int\frac{1}{x}dx}=e^{\ln x}=x$.

To show that $e^{\ln x}=x$, take logs on both side to the base e.

$\ln e^{\ln x}=\ln x \Rightarrow \ln x \ln e=\ln x$ and since $\ln e=1$, LHS = RHS.

Multiplying each term of (1) by x.

$$x\frac{dy}{dx}+y=xe^{-x} \qquad \frac{d}{dx}(xy)=xe^{-x}\Rightarrow d(xy)=xe^{-x}dx$$

intergrating both sides with respect to x

$$\int d(xy)=\int \underset{②}{x}\,\underset{①}{e^{-x}}\,dx \Rightarrow xy=-e^{-x}x-\int -e^{-x}.1\,dx$$

$$xy = -xe^{-x} + \int e^{-x}\mathrm{d}x = -xe^{-x} - e^{-x} + c.$$

If $x = y = 0$, $\quad 0 = -e^{-0} + c \Rightarrow c = 1$

$$xy = -xe^{-x} - e^{-x} + 1$$

$$\boxed{y = -e^{-x} - \frac{1}{x}e^{-x} + \frac{1}{x}}$$

---

5. $(\cos\theta + i\sin\theta)^3 = \cos 3\theta + i\sin 3\theta$

$$= \cos^3\theta + 3\cos^2\theta\,(i\sin\theta) + 3\cos\theta\left(i^2\sin^2\theta\right) + i^3(\sin\theta)^3$$

$$= \left(\cos^3\theta - 3\cos\theta\sin^2\theta\right) + i\left(3\cos^2\theta\sin\theta - \sin^3\theta\right).$$

Equating real and imaginary terms

$$\cos 3\theta = \cos^3\theta - 3\cos\theta\sin^2\theta$$

$$\sin 3\theta = 3\cos^2\theta\sin\theta - \sin^3\theta$$

$$\therefore \tan 3\theta = \frac{\sin 3\theta}{\cos 3\theta} = \frac{3\cos^2\theta\sin\theta - \sin^3\theta}{\cos^3\theta - 3\cos\theta\sin^2\theta}$$

$$= \frac{\cos^3\theta}{\cos^3\theta}\left(3\frac{\sin\theta}{\cos\theta} - \frac{\sin^3\theta}{\cos^3\theta}\right) \times \frac{1}{1 - 3\dfrac{\sin^2\theta}{\cos^2\theta}} = \frac{3\tan\theta - \tan^3\theta}{1 - 3\tan^2\theta}.$$

---

6. Let $\mathrm{f}(a) = a^n$

$$\mathrm{f}'(a) = na^{n-1}$$

$$\mathrm{f}''(a) = n(n-1)a^{n-2}$$

$$\mathrm{f}'''(a) = n(n-1)(n-2)a^{n-3}$$

$$\ldots$$

Using Taylor's theorem which states

$$\mathrm{f}(x + h) = \mathrm{f}(x) + h\,\mathrm{f}'(x) + \frac{h^2}{2!}\mathrm{f}''(x) + \frac{h^3}{3}\mathrm{f}'''(x) + \ldots$$

if $x = a$ and $h = b$

$$(a + b)^n = \mathrm{f}(a) + b\,\mathrm{f}'(a) + \frac{b^2}{2!}\mathrm{f}''(a) + \frac{b^3}{3!}\mathrm{f}'''(a) + \ldots$$

$$= a^n + na^{n-1}b + \frac{n(n-1)}{2}a^{n-2}b^2 + \frac{n(n-1)(n-2)}{3!}a^{n-3}b^3 + \ldots$$

---

7. $z_1 = 3 + i4 \qquad |z_1| = \sqrt{3^2 + 4^2} = 5$

$$\arg z_1 = \tan^{-1}\frac{4}{3} = 53.1^\circ = 0.927^c \text{ to 3 s.f.}$$

$$z_1 = 5\,\underline{/0.927^c} = 5\,(\cos 0.927^c + i \sin 0.927^c)$$

$$z_2 = 5 + i12 \qquad |Z_2| = \sqrt{5^2 + 12^2} = 13$$

$$\arg z_2 = \tan^{-1}\frac{12}{5} = 1.18^c$$

$$z_2 = 13\,\underline{/1.18^c} = 13\,(\cos 1.18^c + i \sin 1.18^c)$$

(i) $z_1 z_2 = 5 \times 13\,\underline{/0.927^c}\,\underline{/1.18^c} = 65\,\underline{/2.10^c}$

$$= 65\,(\cos 2.10^c + i \sin 2.10^c)$$

(ii) $\dfrac{z_1}{z_2} = \dfrac{5\,\underline{/0.927^c}}{13\,\underline{/1.18^c}} = 0.385\,\underline{/-0.249^c} = 0.385\left(\cos 0.249^c - i \sin 0.249^c\right).$

---

8. (i) $(x - 1)(x + 2)(x - 3) < 0$, sketching the graph $f(x) = (x - 1)(x + 2)(x - 3)$

$$f(2) = (2 - 1)\,4\,(2 - 3) = 1 \times 4 \times -1 = -4 < 0$$

$$f(-1) = -2 \times 1 \times -4 = 8 > 0$$

$$f(0) = (-1)(2)(-3) = 6$$

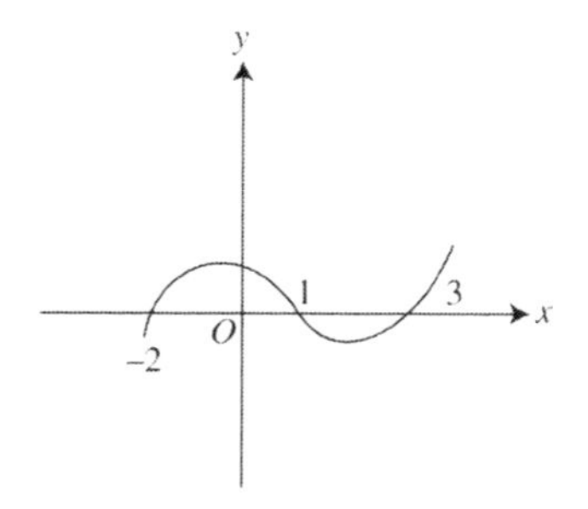

$1 < x < 3 \qquad x < -2.$

(ii) $x(x+1)(x-2) > 0$

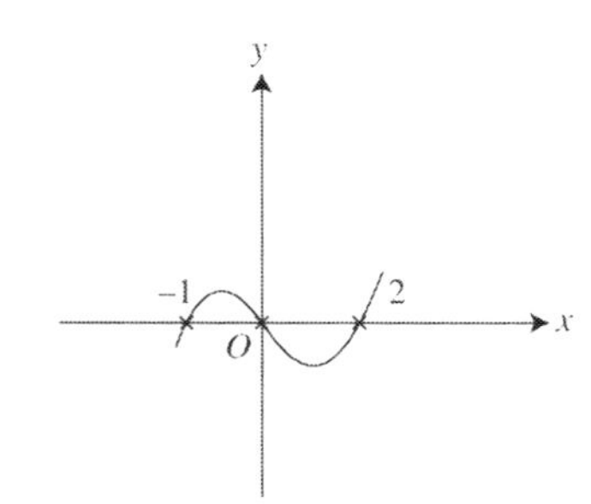

$f(x) = x(x+1)(x-2)$

$f(0) = 0$

$f(1) = 1 \times 2 \times -1 = -2$

$-1 < x < 0 \quad x > 2.$

Examination Test Papers.

GCE Advanced Level.

Further Pure Mathematics FP2

Test Paper 4

Time: 1 hour and 30 minutes.

Instructions and Information

Candidates may use any calculator allowed by the regulations of their Examination Board.

Full marks are awarded for correct answers to ALL questions.

This paper has eight questions.

You can start working with any question and you must label clearly all parts.

1. Convert the following polar coordinates to Cartesian and vice-versa.

(i) $\left(3, \frac{\pi}{4}\right)$, (ii) $\left(2, \frac{\pi}{6}\right)$, (iii) $\left(7, -\frac{\pi}{3}\right)$. **(12)**

---

2. Find the first five terms in the expansion of $\ln(1 + \sin x)$,
using Maclaurin's theorem. **(9)**

---

3. Find the equation of the tangent, which is parallel to the initial line, on the curve $r = \cos\theta$ in the range $0 \leq \theta \leq \frac{\pi}{2}$, you may consider the stationary values of $r\sin\theta$. **(7)**

---

4. Find the general solution $\frac{dx}{d\theta} + (\cot\theta)\,x = \cos\theta$. If $\theta = \frac{\pi}{2}$ for $x = 1$,
find the particular solution. **(12)**

---

5. (a) if $z = \cos\theta + i\sin\theta$ find $\frac{1}{z}$ and hence find:

(i) $z + \frac{1}{z}$ (ii) $z - \frac{1}{z}$ (iii) $z^n + \frac{1}{z^n}$ (iv) $z^n - \frac{1}{z^n}$.

(b) Express $\sin 3\theta$ interms of $\sin\theta$ and $\cos 3\theta$ in terms of $\cos\theta$. **(8)**

---

6. If y = $e^{-2x}\cos 2x$, show that $\frac{d^2y}{dx^2} + 4\frac{dy}{dx} + 8y = 0$.

Hence find the general solution of the second order differential equation. **(10)**

---

7. (a) Express the complex number $z = -1 + i$ in the form $r(\cos\theta + i\sin\theta)$,
$-\pi < \theta \leq \pi$.

(b) Solve the equation $z^3 = -1 + i$, giving the roots in the form $r(\cos\theta + i\sin\theta)$,
$-\pi < \theta \leq \pi$.

(c) Show the roots of $z^3 = -1 + i$ on an Argand diagram. **(10)**

---

8. Solve the inequalities simultaneously

$$|x - 2| < 3$$

$$|2x + 3| \leq 11. \qquad \textbf{(7)}$$

TOTAL FOR PAPER: 75 MARKS

Examination Test Papers.

GCE Advanced Level.

Test Paper 4 Solutions

Further Pure Mathematics FP2

1. Polar to Cartesian.

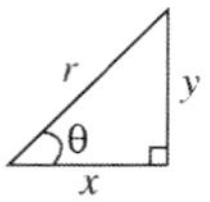

$$\cos\theta = \frac{x}{r} \qquad \sin\theta = \frac{y}{r}$$

$$x = r\cos\theta \qquad y = r\sin\theta$$

(i) $\left(3, \frac{\pi}{4}\right)$ $\quad r = 3 \quad \theta = \frac{\pi}{4}$

$$x = 3\cos\frac{\pi}{4} = \frac{3}{\sqrt{2}}, \qquad y = 3\sin\frac{\pi}{4} = \frac{3}{\sqrt{2}}$$

(ii) $\left(2, \frac{\pi}{6}\right)$ $\quad r = 2 \quad \theta = \frac{\pi}{6}$

$$x = 2\cos\frac{\pi}{6}, \qquad y = 2\sin\frac{\pi}{6}$$

$$x = 2\frac{\sqrt{3}}{2} = \sqrt{3} \qquad y = 2\left(\frac{1}{2}\right) = 1$$

(iii) $\left(7, -\frac{\pi}{3}\right)$ $\quad r = 7 \quad \theta = -\frac{\pi}{3}$

$$x = r\cos\theta = 7\cos\left(-\frac{\pi}{3}\right) = \frac{7}{2}$$

$$y = r\sin\theta = 7\sin\left(-\frac{\pi}{3}\right) = -\frac{7\sqrt{3}}{2}.$$

Cartesian to Polar.

(i) $\left(\frac{3}{\sqrt{2}}, \frac{3}{\sqrt{2}}\right)$

$$x = r\cos\theta \quad \ldots(1)$$

$$y = r\sin\theta \quad \ldots(2)$$

dividing (2) by (1)

$$\frac{y}{x} = \tan\theta = \frac{3\sqrt{2}}{3\sqrt{2}} = 1 \Rightarrow \theta = \frac{\pi}{4}$$

$$r = \sqrt{x^2 + y^2} = \sqrt{\left(\frac{3}{\sqrt{2}}\right)^2 + \left(\frac{3}{\sqrt{2}}\right)^2} = 3 \qquad \therefore \left(\frac{3}{\sqrt{2}}, \frac{3}{\sqrt{2}}\right) \equiv \left(3, \frac{\pi}{4}\right).$$

(ii) $\tan\theta = \dfrac{y}{x} = \dfrac{1}{\sqrt{3}} \qquad \theta = \dfrac{\pi}{6}$

$$r = \sqrt{\left(\sqrt{3}\right)^2 + 1} = 2$$

$$\left(\sqrt{3}, 1\right) \equiv \left(2, \frac{\pi}{6}\right).$$

(iii) $\tan\theta = \dfrac{y}{x} = \dfrac{-7\frac{\sqrt{3}}{2}}{\frac{7}{2}} = -\sqrt{3} \Rightarrow \theta = -\dfrac{\pi}{3}$

$$r = \sqrt{\left(\frac{7}{2}\right)^2 + \left(\frac{-7\sqrt{3}}{2}\right)^2} = 7$$

$$\therefore \left(\frac{7}{2}, \frac{-7\sqrt{3}}{2}\right) \equiv \left(7, -\frac{\pi}{3}\right).$$

---

2. $\mathrm{f}(x) = \ln(1 + \sin x) \qquad \mathrm{f}(0) = 0$

$$\mathrm{f}'(x) = \frac{1}{1 + \sin x} \times \cos x = \frac{\cos x}{1 + \sin x} \qquad \mathrm{f}'(0) = 1$$

$$\mathrm{f}''(x) = \frac{(-\sin x)(1 + \sin x) - \cos x \cos x}{(1 + \sin x)^2}$$

$$= \frac{-\sin x - \sin^2 x - \cos^2 x}{(1 + \sin x)^2} = -\frac{1 + \sin x}{(1 + \sin x)^2} = -\frac{1}{1 + \sin x} \qquad \mathrm{f}''(0) = -1$$

$$\mathrm{f}'''(x) = \frac{\cos x}{(1 + \sin x)^2} \qquad \mathrm{f}'''(0) = 1$$

$$\mathrm{f}^{(\mathrm{iv})}(x) = \frac{(-\sin x)(1 + \sin x)^2 - 2\cos x\,(1 + \sin x)\cos x}{(1 + \sin x)^4}$$

$$= -\frac{\sin x + \sin^2 x + 2\cos^2 x}{(1 + \sin x)^3} \Rightarrow \mathrm{f}^{(\mathrm{iv})}(0) = -2$$

$$f^{(v)}(x) = \frac{\left\{(-\cos x - 2\sin x\cos x + 4\cos x\sin x)(1+\sin x)^3 + (\sin x + \sin^2 x + 2\cos^2 x)\,3\,(1+\sin x)^2\cos x\right\}}{(1+\sin x)^6}$$

$$f^{(v)}(0) = \frac{-1+6}{1} = 5 \qquad \therefore \ln(1+\sin x) = x - \frac{x^2}{2} + \frac{x^3}{6} - \frac{x^4}{12} + \frac{x^5}{24}.$$

---

3. The derivative of $r\sin\theta$ is $\quad \frac{d}{d\theta}(r\sin\theta) = \frac{dr}{d\theta}\sin\theta + r\cos\theta.$

Note that $r\sin\theta$ in cartesian coordinates is equal to y.

For stationary values $\frac{dr}{d\theta}\sin\theta + r\cos\theta = 0. \quad \ldots(1)$

The curve is $r = \cos\theta$ then $\frac{dr}{d\theta} = -\sin\theta$ from(1) $(-\sin\theta)\times(\sin\theta) + \cos\theta\cos\theta = 0$

$\cos^2\theta - \sin^2\theta = 0 \qquad \cos 2\theta = 0 = \cos\frac{\pi}{2} \Rightarrow \theta = \frac{\pi}{4}.$

The coordinates of P are $r = \cos\theta = \cos\frac{\pi}{4} = \frac{1}{\sqrt{2}}$ and $\theta = \frac{\pi}{4}$, $P\left(\frac{1}{\sqrt{2}}, \frac{\pi}{4}\right)$.

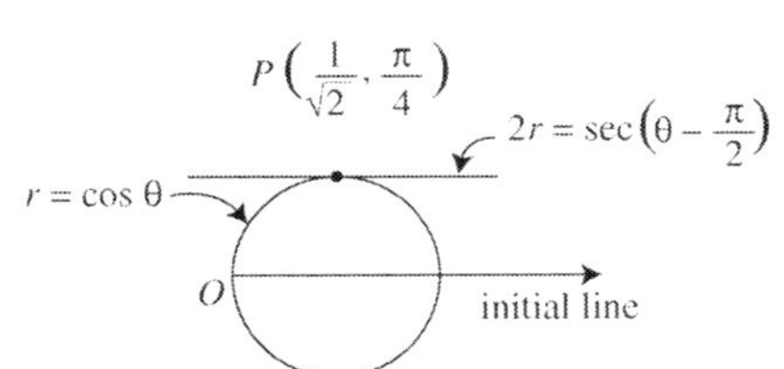

The equation of the tangent at $P\left(\frac{1}{\sqrt{2}}, \frac{\pi}{4}\right)$

$r\sin\theta = \cos\theta\sin\theta = \cos\frac{\pi}{4}\sin\frac{\pi}{4} = \frac{1}{\sqrt{2}}\frac{1}{\sqrt{2}} = \frac{1}{2} \quad$ since $r = \cos\theta$

$r\sin\theta = \frac{1}{2} \Rightarrow r = \frac{1}{2\sin\theta} = \frac{1}{2}\operatorname{cosec}\theta$

$r = \frac{1}{2}\operatorname{cosec}\theta \qquad 2r = \sec\left(\theta - \frac{\pi}{2}\right).$

---

4. $\frac{dx}{d\theta} + x\cot\theta = \cos\theta \qquad \ldots(1)$

this is of the form $\frac{dx}{d\theta} + Px = Q$ where P and Q are the functions of θ.

I.F. $= e^{\int \cot\theta\, d\theta} = e^{\ln\sin\theta} = \sin\theta$

$\sin\theta\frac{dx}{d\theta} + x\sin\theta\cot\theta = \sin\theta\cos\theta$

If (1) is multiplied by $\sin\theta$ on both sides

$$\frac{d}{d\theta}(x\sin\theta) = \sin\theta\cos\theta = \frac{\sin 2\theta}{2}$$

$$\therefore d(x\sin\theta) = \frac{1}{2}\sin 2\theta\, d\theta \qquad \ldots(2)$$

Integrating both sides of (2)

$$\int d(x\sin\theta) = \int \frac{1}{2}\sin 2\theta\, d\theta$$

$$x\sin\theta = -\frac{\cos 2\theta}{4} + c \text{ is the general solution.}$$

If $\theta = \frac{\pi}{2}$ for $x = 1$.

$$1\sin\frac{\pi}{2} = -\frac{\cos\pi}{4} + c \Rightarrow c = 1 + \frac{\cos\pi}{4}$$

$$c = 1 - \frac{1}{4} = \frac{3}{4} \Rightarrow \boxed{c = \frac{3}{4}}$$

$$x\sin\theta = \frac{-\cos 2\theta}{4} + \frac{3}{4}$$

$$\therefore x = -\frac{\cos 2\theta}{4\sin\theta} + \frac{3}{4\sin\theta}$$

$$\boxed{x = \frac{3}{4}\operatorname{cosec}\theta - \frac{1}{4}\cos 2\theta\operatorname{cosec}\theta}$$

---

5. (a) $z = \cos\theta + i\sin\theta \qquad \frac{1}{z} = \cos\theta - i\sin\theta$

(i) $z + \frac{1}{z} = 2\cos\theta \qquad$ (ii) $z - \frac{1}{z} = 2i\sin\theta$

(iii) $$z^n = \cos n\theta + i\sin n\theta$$

$$z^{-n} = \cos n\theta - i\sin n\theta$$

$$z^n + \frac{1}{z^n} = 2\cos n\theta$$

$$z^n - \frac{1}{z^n} = 2i\sin n\theta.$$

(b) $(\cos\theta + i\sin\theta)^3 = \cos 3\theta + i\sin 3\theta$

$$= \cos^3\theta + i^3\sin^3\theta + 3\cos^2\theta\, i\sin\theta + 3\cos\theta\, i^2\sin^2\theta$$

$$= \cos^3\theta - i\sin^3\theta + 3i\cos^2\theta\sin\theta - 3\cos\theta\sin^2\theta.$$

Equating real and imaginary terms

$$\cos 3\theta = \cos^3\theta - 3\cos\theta\sin^2\theta = \cos^3\theta - 3\cos\theta\left(1-\cos^2\theta\right)$$

$$= \cos^3\theta - 3\cos\theta + 3\cos^3\theta = 4\cos^3\theta - 3\cos\theta$$

$$\sin 3\theta = -\sin^3\theta + 3\cos^2\theta\sin\theta = -\sin^3\theta + 3\left(1-\sin^2\theta\right)\sin\theta$$

$$= 3\sin\theta - 4\sin^3\theta.$$

---

6. If $y = e^{-2x}\cos 2x \qquad \ldots (1)$

$$\frac{dy}{dx} = -2e^{-2x}\cos 2x + e^{-2x}\left[-2\sin 2x\right]$$

$$= -2e^{-2x}\cos 2x - 2e^{-2x}\sin 2x \qquad \ldots (2)$$

$$\frac{d^2y}{dx^2} = 4e^{-2x}\cos 2x + 4e^{-2x}\sin 2x + 4e^{-2x}\sin 2x - 4e^{-2x}\cos 2x$$

$$= 8e^{-2x}\sin 2x \qquad \ldots (3)$$

Substituting (1), (2) and (3)

in $\frac{d^2y}{dx^2} + 4\frac{dy}{dx} + 8y = 0$, we have

L.H.S. $= 8e^{-2x}\sin 2x - 8e^{-2x}\cos 2x - 8e^{-2x}\sin 2x + 8e^{-2x}\cos 2x = 0$

$\therefore$ L.H.S. = R.H.S.

$$\frac{d^2y}{dx^2} + 4\frac{dy}{dx} + 8y = 0$$

$m^2 + 4m + 8 = 0$ the auxiliary equation

$$m = \frac{-4 \pm \sqrt{16-32}}{2} = -2 \pm 2i.$$

The general solution $\qquad y = e^{-2x}\left[A\cos 2x + B\sin 2x\right].$

---

7. (a) $z = -1 + i \quad |z| = \sqrt{(-1)^2 + 1^2} = \sqrt{2}$

$$z = \sqrt{2} \angle \tfrac{3\pi}{4} = \sqrt{2}\left(\cos\frac{3\pi}{4} + i\sin\frac{3\pi}{4}\right).$$

(b) $z^3 = -1 + i, \quad z = \left[\sqrt{2}\left(\cos\frac{3\pi}{4} + i\sin\frac{3\pi}{4}\right)\right]^{\frac{1}{3}}$

$$z = 2^{\frac{1}{6}} \cos\left(\frac{\frac{3\pi}{4} + 2k\pi}{3} + i\sin\frac{\frac{3\pi}{4} + 2k\pi}{3}\right)$$

where $k = 0, \quad \pm 1$

$z_1 = 2^{\frac{1}{6}} \angle \frac{\pi}{4} \quad z_2 = 2^{\frac{1}{6}} \angle \frac{11\pi}{12} \quad z_3 = 2^{\frac{1}{6}} \angle -\frac{5\pi}{12}.$

(c)

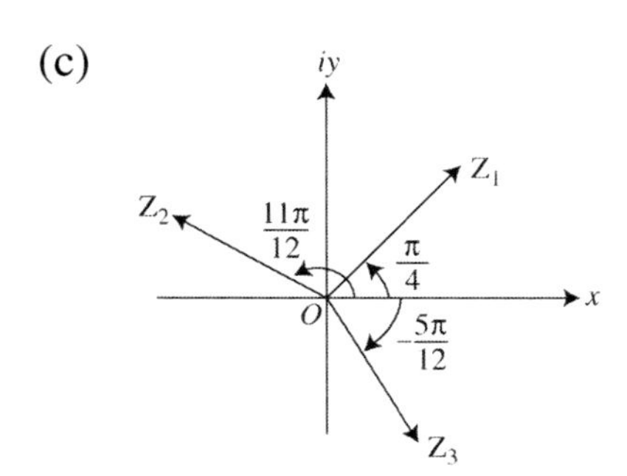

---

8. $|x - 2| < 3$

squaring up both sides

$(x - 2)^2 - 3^2 < 0$

$(x - 2 - 3)(x - 2 + 3) < 0$

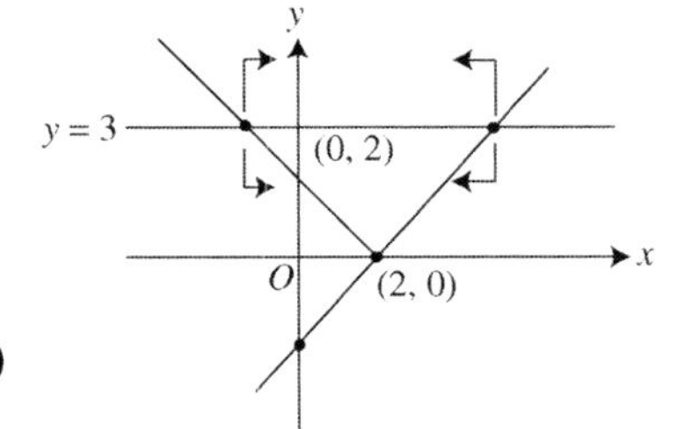

$y = x - 2$

$x = 0 \quad y = -2$

$y = 0 \quad x = 2$

$|y| = |x - 2|$

$(x - 5)(x + 1) < 0 \qquad -1 < x < 5$

$|2x + 3| \le 11 \quad (2x + 3)^2 - 11^2 \le 0 \quad (2x + 3 - 11)(2x + 3 + 11) \le 0$

$2(x - 4)2(x + 7) \le 0 \qquad -7 \le x \le 4 \qquad \therefore \boxed{-1 < x < 4}.$

Examination Test Papers.

GCE Advanced Level.

Further Pure Mathematics FP2

Test Paper 5

Time: 1 hour and 30 minutes.

Instructions and Information

Candidates may use any calculator allowed by the regulations of their Examination Board.

Full marks are awarded for correct answers to ALL questions.

This paper has eight questions.

You can start working with any question and you must label clearly all parts.

1. (a) Sketch the following polar curves:

(i) $r = 1 + \cos\theta$ (ii) $r = 1 - \cos\theta$.

(b) Draw the curve $r = 1 + 2\cos\theta$, making a table of values at 30° intervals up to 360°.

(c) Calculate the area contained in the inner loop. **(13)**

---

2. Expand $\sin(x + h)$ in powers of h by Taylor's Theorem, and deduce the value of $\sin 31^\circ$ correct to 5 decimel places. **(6)**

---

3. By considering the stationary values of $5\cos\theta$, find the polar coordinates of the points on the curve $r = 5\sin\theta$ where the tangents are perpendicular to the initial line. **(6)**

---

4. Solve the differential equation $\frac{\mathrm{d}y}{\mathrm{d}x} + y\tan x = \cos x$ using the integrating factor method.

If $x = \frac{\pi}{4}$, $y = 1$, find y in terms of x. **(10)**

---

5. Use Maclaurin's theorem to show

$$e^{\theta} = 1 + \frac{\theta}{1!} + \frac{\theta^2}{2!} + \ldots.$$

$$e^{i\theta} = \left(1 - \frac{\theta^2}{2!} + \frac{\theta^4}{4!} - \ldots\right) + i\left(\theta - \frac{\theta^3}{3!} + \frac{\theta^5}{5!} - \ldots\right)$$

$$= \cos\theta + i\sin\theta \qquad \text{Euler's formula}$$

where θ is expressed in radians. **(6)**

---

6. Find the first five terms in the expansion of $\ln(1 + \sin x)$, using Maclaurin's theorem. **(6)**

---

7. (a) Express the complex number $z = 1 - i$ in the form $r(\cos\theta + i\sin\theta)$, $-\pi < \theta \le \pi$.

(b) Solve the equation $z^5 = 1 - i$, giving the roots in the form

$r(\cos\theta + i\sin\theta)$, $-\pi < \theta \le \pi$.

(c) Show these roots on an Argand diagram. **(9)**

---

8. (a) Express $\dfrac{1}{(r+1)(r+3)}$ in partial fraction. **(4)**

(b) Hence prove by method of differences that

$$\sum_{r=11}^{n} \frac{1}{(r+1)(r+3)} = \frac{n(5n+13)}{12(n+2)(n+3)}.$$ **(10)**

(c) Show that the rational value of

$$\sum_{r=11}^{20} \frac{1}{(r+1)(r+3)} = \frac{2815}{78936}.$$ **(5)**

TOTAL FOR PAPER: 75 MARKS

Examination Test Papers.

GCE Advanced Level.

Test Paper 5 Solutions

Further Pure Mathematics FP2

1. (a) (i) $r = 1 + \cos\theta$ (ii) $r = 1 - \cos\theta$

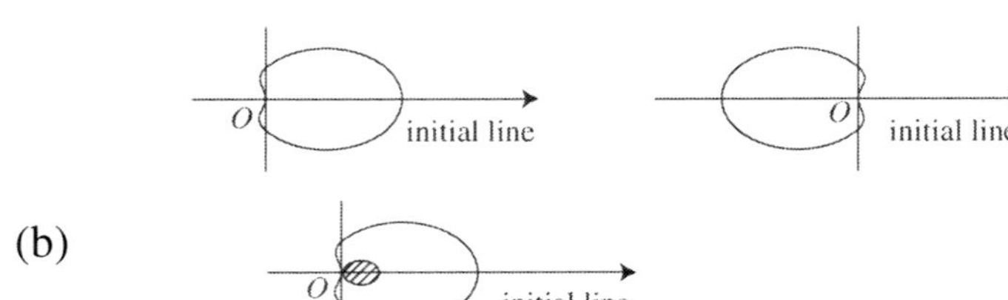

(b)

$\theta°$	0	30	60	90	120	150	180	210	240	270	300	330	360
$\cos\theta$	1	0.866	0.5	0	−0.5	−0.866	−1	−0.866	−0.5	0	0.5	0.866	1
$2\cos\theta$	2	1.732	1	0	−1	−1.732	−2	−1.732	−1	0	1	1.732	2
$r = 1 + 2\cos\theta$	3	2.732	2	1	0	−0.732	−1	−0.732	0	1	2	2.732	3

The limits of the integration are $\frac{2\pi}{3}$ and π.

The area contained in the inner loop

$$= 2 \times \frac{1}{2}\int_{\frac{2\pi}{3}}^{\pi} (1 + 2\cos\theta)^2 d\theta$$

$$= \int_{\frac{2\pi}{3}}^{\pi} \left(1 + 4\cos\theta + 4\cos^2\theta\right) d\theta$$

$$\cos^2\theta = \frac{\cos 2\theta + 1}{2}$$

$$= \int_{\frac{2\pi}{3}}^{\pi} (1 + 4\cos\theta) + 2(\cos 2\theta + 1)\, d\theta$$

$$= \left[\theta + \sin 2\theta + 2\theta + 4\sin\theta\right]_{\frac{2\pi}{3}}^{\pi}$$

$$= 3\pi - \frac{2\pi}{3}(3) + \sin 2\pi - \sin\frac{4\pi}{3} + 4\sin\pi - 4\sin\frac{2\pi}{3}$$

$$= \pi + \frac{\sqrt{3}}{2} - \frac{4\sqrt{3}}{2} = \pi - \frac{3\sqrt{3}}{2}.$$

2. Let $f(x+h) = \sin(x+h)$

$f(x) = \sin x \quad f'(x) = \cos x \quad f''(x) = -\sin x.$

Taylor's Theorem gives

$$\sin(x+h) = \sin x + h\cos x + h^2\frac{(-\sin x)}{2!} + h^3\frac{(-\cos x)}{3!} + \cdots$$

$$\sin(x+h) = \sin x + h\cos x - \frac{h^2}{2!}\sin x - \frac{h^3}{3!}\cos x + \cdots$$

Let $x = 30° = \frac{\pi}{6}$

$$\sin\left(30° + h\right) = \sin 30° + h\cos 30° - \frac{h^2}{2!}\sin 30° - \frac{h^3}{3!}\cos 30° + \cdots$$

$$= \frac{1}{2} + h\frac{\sqrt{3}}{2} - \frac{h^2}{2}\cdot\frac{1}{2} - \frac{1}{6}h^3\frac{\sqrt{3}}{2} + \cdots$$

If now $h = 1° = 0.01745$, $h^2 = 0.000304$, and $h^3 = 0.000005$

$\sin 31° \approx 0.5 + 0.866 \times 0.01745 - 0.25 \times 0.000304$

$\sin 31° \approx 0.51504.$

From the calculator $\sin 31° = 0.515038 \approx 0.51504$ correct to 5 decimal places.

---

3. Differentiating $r\cos\theta$ with respect to θ

$$\frac{d}{d\theta}(r\cos\theta) = \frac{dr}{d\theta}\cos\theta + r(-\sin\theta) = \frac{dr}{d\theta}\cos\theta - r\sin\theta.$$

For stationary values

$$\frac{dr}{d\theta}\cos\theta - r\sin\theta = 0 \quad \ldots(1)$$

$r = 5\sin\theta$, $\frac{dr}{d\theta} = 5\cos\theta$ and substituting in (1)

$$5\cos\theta\cos\theta - 5\sin\theta\sin\theta = 5\cos 2\theta = 0 = \cos\frac{\pi}{2} = \cos\frac{3\pi}{2}$$

$$\therefore \theta = \frac{\pi}{4} \text{ or } \theta = \frac{3\pi}{4}.$$

The coordinates at P and Q are $P\left(\frac{5}{\sqrt{2}}, \frac{\pi}{4}\right)$, $Q\left(\frac{5}{\sqrt{2}}, \frac{3\pi}{4}\right)$ and the corresponding

tangents are $r = \frac{5}{2}\sec\theta$ and $r = -\frac{5}{2}\sec\theta$

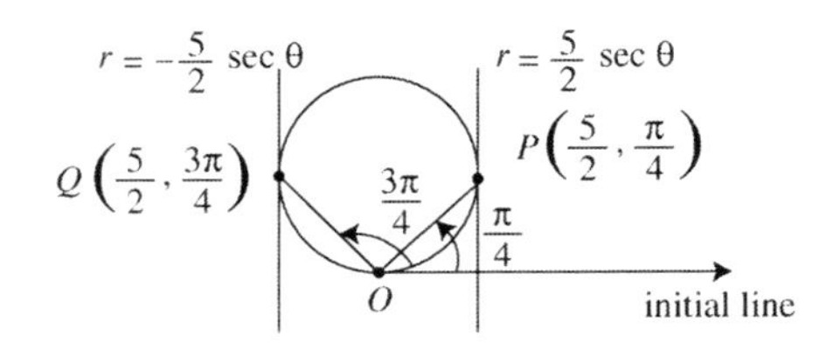

---

4. $\frac{\mathrm{d}y}{\mathrm{d}x} + y\tan x = \cos x \qquad \ldots(1)$

This equation is of the form $\frac{\mathrm{d}y}{\mathrm{d}x} + Py = Q$

where P and Q are functions of x, $\qquad P = \tan x$ and $Q = \cos x$.

The intergrating factor, I.F. $= e^{\int \tan x \, \mathrm{d}x}$

$$\text{I.F.} = e^{\int \frac{\sin x}{\cos x}\mathrm{d}x} = e^{-\int \frac{\mathrm{d}(\cos x)}{\cos x}} = e^{-\ln|\cos x|} = e^{\ln|\sec x|} = \sec x.$$

Multiplying each side of the equation (1) with the I.F.

$$\sec x \frac{\mathrm{d}y}{\mathrm{d}x} + \sec x \; y\tan x = \sec x \; \cos x \quad \frac{\mathrm{d}}{\mathrm{d}x}(y\sec x) = 1$$

$$y\sec x = \int \mathrm{d}x = x + c$$

the general solution

$$y = \frac{x}{\sec x} + \frac{c}{\sec x} = x\cos x + c\cos x \quad \therefore y = x\cos x + c\cos x \quad \text{general solution}$$

when $x = \frac{\pi}{4}$, $y = 1 \quad y = \frac{\pi}{4}\cos\frac{\pi}{4} + c\cos\frac{\pi}{4}$

$$1 = \frac{\pi}{4}\frac{1}{\sqrt{2}} + \frac{c}{\sqrt{2}} \Rightarrow c = \sqrt{2} - \frac{\pi}{4}$$

$$\therefore y = x\cos x + \left(\sqrt{2} - \frac{\pi}{4}\right)\cos x$$

$$\boxed{y = \left(x + \sqrt{2} - \frac{\pi}{4}\right)\cos x}$$ the particular solution.

---

5. $f(x) = f(0) + f'(0)\,\dfrac{x}{1!} + f''(0)\,\dfrac{x^2}{2!} + \cdots$

Maclaurin's theorem

$f(\theta) = e^{\theta} \quad f(0) = 1, \quad f'(\theta) = e^{\theta} \quad f'(0) = 1, \quad f''(\theta) = e^{\theta} \quad f''(0) = 1$

$$\therefore e^{\theta} = 1 + \frac{\theta}{1!} + \frac{\theta}{2!} + \cdots$$

$f(\theta) = e^{i\theta} \quad f(0) = 1, \quad f'(\theta) = ie^{i\theta} \quad f'(0) = i \quad f''(\theta) = i^2 e^{i\theta} \quad f''(0) = -1$

$$e^{i\theta} = 1 + i\frac{\theta}{1} - \frac{\theta^2}{2!} + \frac{i^3\theta^3}{3!} + \frac{i^4\theta^4}{4!}$$

$$= \left(1 - \frac{\theta^2}{2!} + \frac{\theta^4}{4!} - \cdots\right) + i\left(\theta - \frac{\theta^3}{3!} + \frac{\theta^5}{5!} - \cdots\right) = \cos\theta + i\sin\theta$$

θ is expressed in radians.

---

6. $f(x) = \ln(1 + \sin x) \qquad f(0) = 0 \quad f'(x) = \dfrac{\cos x}{1 + \sin x} \qquad f'(0) = 0$

$$f''(x) = \frac{(-\sin x)(1 + \sin x) - \cos^2 x}{(1 + \sin x)^2} = \frac{-\sin x - \sin^2 x - \cos^2 x}{(1 + \sin x)^2}$$

$$= -\frac{1 + \sin x}{(1 + \sin x)^2} = -\frac{1}{1 + \sin x} \qquad f''(0) = 1$$

$$f'''(x) = \frac{\cos x}{(1 + \sin x)^2} \qquad f'''(0) = 1$$

$$f^{(iv)}(x) = \frac{-\sin x - \sin^2 x - 2\cos^2 x}{(1 + \sin x)^3} \qquad f^{(iv)}(0) = -2$$

$f^{(v)}(0) = 5$

$$\therefore \ln(1 + \sin x) = x - \frac{x^2}{2} + \frac{x^3}{6} - \frac{x^4}{12} + \frac{x^5}{24}.$$

---

7. (a) $z = 1 - i \quad |z| = \sqrt{2} \quad \arg z = \angle -\frac{\pi}{4}$

$$z = \sqrt{2} \angle -\frac{\pi}{4} = \sqrt{2}\left(\cos\frac{\pi}{4} - i\sin\frac{\pi}{4}\right).$$

(b) $z^5 = \sqrt{2}\left(\cos\frac{\pi}{4} - i\sin\frac{\pi}{4}\right)$

$$z = 2^{\frac{1}{10}} \angle \frac{-\frac{\pi}{4} + 2k\pi}{5} \qquad \text{where } k = 0, \pm 1, \pm 2$$

$$z_1 = 2^{\frac{1}{10}} \angle -\frac{\pi}{20} \qquad z_2 = 2^{\frac{1}{10}} \angle \frac{7\pi}{20} \qquad z_3 = 2^{\frac{1}{10}} \angle -\frac{9\pi}{20}$$

$$z_4 = 2^{\frac{1}{10}} \angle \frac{15\pi}{20} \qquad z_5 = 2^{\frac{1}{10}} \angle -\frac{17\pi}{20}.$$

(c)

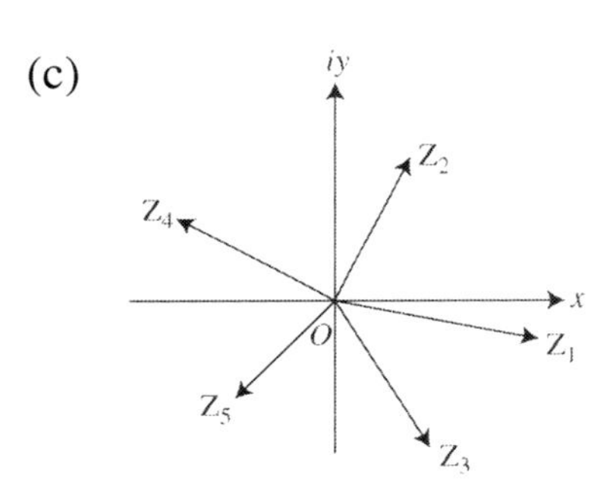

---

8. (a) $$\frac{1}{(r+1)(r+3)} \equiv \frac{A}{r+1} + \frac{B}{r+3}$$

$$1 \equiv A(r+3) + B(r+1)$$

if $r = -1$

$$1 = 2A \Rightarrow A = \frac{1}{2}$$

if $r = -3$

$$1 = -2B \Rightarrow B = -\frac{1}{2}$$

$$\therefore \frac{1}{(r+1)(r+3)} \equiv \frac{\frac{1}{2}}{r+1} - \frac{\frac{1}{2}}{r+3}.$$

(b) Using method of differences

$$\text{when } r = 1 \qquad \frac{1}{2\times 2} - \frac{1}{2\times 4}$$

$$r = 2 \qquad \frac{1}{2\times 3} - \frac{1}{2\times 5}$$

$$r = 3 \qquad \frac{1}{2\times 4} - \frac{1}{2\times 6}$$

$$\ldots$$

$$r = n-1 \qquad \frac{1}{2\times n} - \frac{1}{2\times (n+2)}$$

$$r = n \qquad \frac{1}{2\times (n+1)} - \frac{1}{2\times (n+3)}$$

$$\therefore \sum_{r=1}^{n} \frac{1}{(r+1)(r+3)} = \frac{1}{4} + \frac{1}{6} - \frac{1}{2(n+2)} - \frac{1}{2\times (n+3)}$$

$$= \frac{5}{12} - \frac{1}{2}\left[\frac{n+3+n+2}{(n+2)(n+3)}\right]$$

$$= \frac{5}{12} - \frac{(2n+5)}{2(n+2)(n+3)}$$

$$= \frac{5(n+2)(n+3) - 6(2n+5)}{12(n+2)(n+3)}$$

$$= \frac{5n^2 + 25n + 30 - 12n - 30}{12(n+2)(n+3)}$$

$$= \frac{5n^2 + 13n}{12(n+2)(n+3)} = \frac{n(5n+13)}{12(n+2)(n+3)}$$

$$= \frac{n(5n+13)}{12(n+2)(n+3)}.$$

(c) $$\sum_{r=11}^{20} \frac{1}{(r+1)(r+3)} = \sum_{r=1}^{20} \frac{1}{(r+1)(r+3)} - \sum_{r=1}^{10} \frac{1}{(r+1)(r+3)}$$

$$= \frac{20(5 \times 20 + 13)}{12(20+2)(20+3)} - \frac{10(5 \times 10 + 13)}{12(10+2)(10+3)}$$

$$= \frac{20 \times 113}{12 \times 22 \times 23} - \frac{10 \times 63}{12 \times 12 \times 13}$$

$$= \frac{10}{12}\left(\frac{2 \times 113}{22 \times 23} - \frac{63}{12 \times 13}\right)$$

$$= \frac{5}{6 \times 2}\left(\frac{2 \times 113}{253} - \frac{63}{78}\right)$$

$$= \frac{5}{12}\left(\frac{226}{253} - \frac{21}{26}\right)$$

$$= \frac{5}{12}\left(\frac{5876 - 5313}{26 \times 253}\right) = \frac{5}{12} \times \frac{563}{6578}$$

$$= \frac{2815}{78936}.$$

Examination Test Papers.

GCE Advanced Level.

Further Pure Mathematics FP2

Test Paper 6

Time: 1 hour and 30 minutes.

Instructions and Information

Candidates may use any calculator allowed by the regulations of their Examination Board.

Full marks are awarded for correct answers to ALL questions.

This paper has eight questions.

You can start working with any question and you must label clearly all parts.

1. Draw the curve $r = 1 - 2\sin\theta$ making a table of values at $30°$ intervals up to $360°$. Calculate the area contained in the inner loop.

---

2. Find the solution, in ascending powers of x up to and including the term of the differential equation $\frac{d^2y}{dx^2} + (x+1)\frac{dy}{dx} + y = 0$ given that, when $x = 0$, $y = 1$ and $\frac{dy}{dx} = 2$.

---

3. Sketch the following graphs:

a) (i) $\theta = \frac{\pi}{4}$ (ii) $\theta = \frac{3\pi}{4}$ (iii) $\theta = -\frac{\pi}{4}$ (iv) $\theta = \frac{5\pi}{4}$.

b) (i) $r = \sec\theta$ (ii) $r = 2\sec\theta$ (iii) $r = 3\sec\theta$.

c) (i) $r = \operatorname{cosec}\theta$ (ii) $r = 2\operatorname{cosec}\theta$ (iii) $r = 3\operatorname{cosec}\theta$.

Explain and compare with the corresponding Cartesian expressions and show that $r = \operatorname{cosec}\theta = \sec\left(\theta - \frac{\pi}{2}\right)$.

---

4. Solve $6\frac{d^2y}{dx^2} - \frac{dy}{dx} - y = 10\sin 2x$.

---

5. (a) Find the square roots of $3 + 4i$.

(b) Determine the roots of the cubic equation $z^3 - 1 = 0$.

---

6. Use Maclaurin's theorem to show that $f(x) = \sin^2 x \approx x^2 - \frac{1}{3}x^4 + \frac{2}{45}x$

hence show that $\cos^2 x \approx 1 - x^2 + \frac{1}{3}x^4 - \frac{2}{45}x^6$ and

$$\cos 2x \approx 1 - 2x^2 + \frac{2}{3}x^4 - \frac{4}{45}x^6.$$

---

7. (a) Express the complex number $z = -3 - 4i$ in the polar form $0° < \theta$

(b) Solve the equation $z^3 = -3 - 4i$ in the polar form $0° < \theta < 360°$.

(c) Show these roots on an Argand diagram. **(9)**

---

8. (a) Express $\dfrac{1}{(r+3)(r+5)}$ in partial fractions. **(4)**

(b) Hence find the sum of the series

$\displaystyle\sum_{r=1}^{n} \frac{1}{(r+3)(r+5)}$ using the method of differences. **(5)**

TOTAL FOR PAPER: 75 MARKS

Examination Test Papers.

GCE Advanced Level.

Test Paper 6 Solutions

Further Pure Mathematics FP2

1.

$\theta°$	0	30	60	90	120	150	180	210	240	270	300	330	360
$\sin\theta$	0	0.5	0.866	1	0.866	0.5	0	−0.5	−0.866	−1	−0.866	−0.5	0
$-2\sin\theta$	0	−1	−0.732	−2	−1.732	−1	0	+1	+1.732	+2	+1.732	+1	0
$r=1-2\sin\theta$	1	0	−0.732	−1	−0.732	0	1	2	2.732	3	2.732	2	1

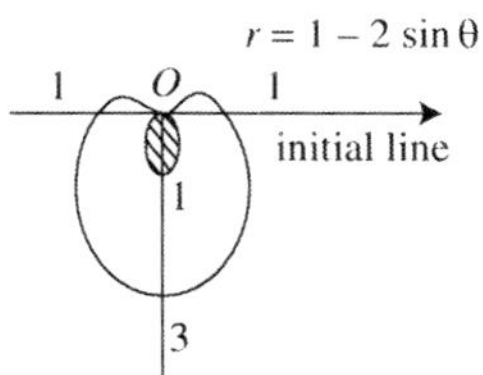

$$2 \times \frac{1}{2}\int_{\frac{\pi}{6}}^{2} r^2\,d\theta = \int_{\frac{\pi}{6}}^{\frac{\pi}{2}} (1-2\sin\theta)^2\,d\theta = \int_{\frac{\pi}{6}}^{\frac{\pi}{2}} \left(1-4\sin\theta+4\sin^2\theta\right)d\theta$$

$$\cos 2\theta = \cos^2\theta - \sin^2\theta = 1 - 2\sin^2\theta$$

$$\sin^2\theta = \frac{1-\cos 2\theta}{2}$$

$$= \int_{\frac{\pi}{6}}^{\frac{\pi}{2}} \left[1 - 4\sin\theta + 4 \times \frac{1-\cos 2\theta}{2}\right] d\theta$$

$$= \left[\theta + 4\cos\theta + 2\left(\theta - \frac{\sin 2\theta}{2}\right)\right]_{\frac{\pi}{6}}^{\frac{\pi}{2}}$$

$$= \left[3\theta + 4\cos\theta - \sin 2\theta\right]_{\frac{\pi}{6}}^{\frac{\pi}{2}}$$

$$= \frac{3\pi}{2} + 0 - 0 - \frac{3\pi}{6} - \frac{4\sqrt{3}}{2} + \frac{\sqrt{3}}{2}$$

$$= \left(\pi - \frac{3}{2}\sqrt{3}\right) \text{ s.u.}$$

---

2. $$\frac{d^2y}{dx^2} + (x+1)\frac{dy}{dx} + y = 0 \qquad \ldots(1)$$

$$\left(\frac{d^2y}{dx^2}\right)_0 + \left(x_0+1\right)\left(\frac{dy}{dx}\right)_0 + y_0 = 0$$

$$x_0 = 0,\ y_0 = 1,\quad \left(\frac{dy}{dx}\right)_0 = 2$$

$$\left(\frac{d^2y}{dx^2}\right)_0 + (2) + 1 = 0 \qquad \left(\frac{d^2y}{dx^2}\right)_0 = -3.$$

Differentiating equation (1) with respect to x

$$\frac{d^3y}{dx^3} + \frac{dy}{dx} + (x+1)\frac{d^2y}{dx^2} + \frac{dy}{dx} = 0$$

$$\left(\frac{d^3y}{dx^3}\right)_0 + \left(\frac{dy}{dx}\right)_0 + (x_0+1)\left(\frac{d^2y}{dx^2}\right)_0 + \left(\frac{dy}{dx}\right)_0 = 0$$

$$\left(\frac{d^3y}{dx^3}\right)_0 + 2 + (0+1)(-3) + 2 = 0 \Rightarrow \left(\frac{d^3y}{dx^3}\right)_0 = -1$$

$$y \approx y_0 + \left(\frac{dy}{dx}\right)_0 (x - x_0) + \left(\frac{d^2y}{dx^2}\right)_0 \frac{(x-x_0)^2}{2!} + \cdots$$

$$y \approx 1 + 2x + (-3)\frac{x^2}{2} + (-1)\frac{x^3}{6}$$

$$y \approx 1 + 2x - \frac{3}{2}x^2 - \frac{1}{6}x^3.$$

---

3. (a) (i) $\theta = \frac{\pi}{4}$; $\frac{\pi}{4}$; O; initial line

(ii)

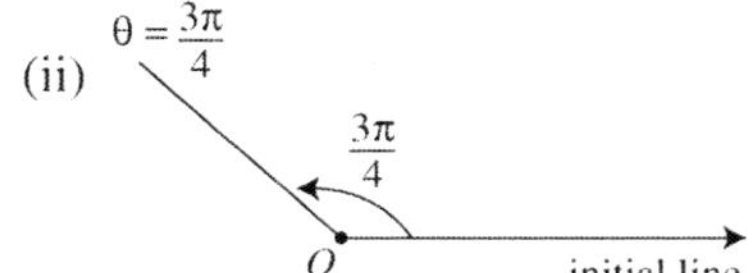

(iii) O; initial line; $-\frac{\pi}{4}$; $\theta = -\frac{\pi}{4}$

(iv)

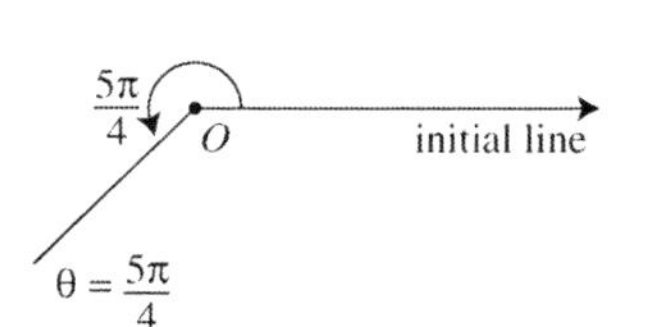

These are half-lines

(i) $y = x$ half of this line

(ii) $y = -x$ half of this line

(iii) $y = -x$ half of this line

(iv) $y = x$ half of this line.

(b)

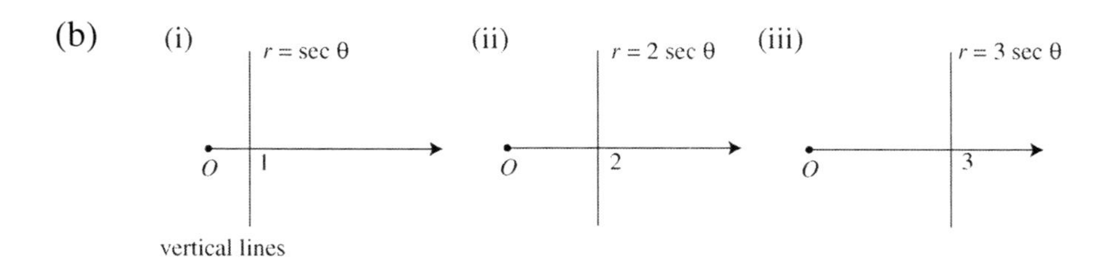

(c)

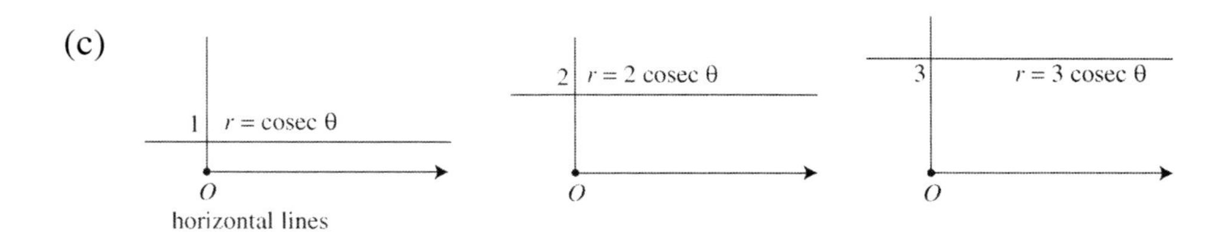

$$r\cos\theta = x \Rightarrow r = x\sec\theta, \quad \text{if } x = 1, r = \sec\theta$$

$$r\sin\theta = y \Rightarrow r = y\operatorname{cosec}\theta, \quad \text{if } y = 1, r = \operatorname{cosec}\theta$$

$$\sec\left(\theta - \frac{\pi}{2}\right) = \frac{1}{\cos\left(\theta - \frac{\pi}{2}\right)} = \frac{1}{\cos\theta\cos\frac{\pi}{2} + \sin\theta\sin\frac{\pi}{2}}$$

$$= \operatorname{cosec}\theta.$$

---

4. $6\frac{d^2y}{dx^2} - \frac{dy}{dx} - y = 10\sin 2x.$

The auxiliary equation is $6m^2 - m - 1 = 0$

which factorises to $(2m - 1)(3m + 1) = 0$,

$m_1 = \frac{1}{2}$ and $m_2 = -\frac{1}{3}$,

The C.F. solution

$$y = Ae^{\frac{1}{2}x} + Be^{-\frac{1}{3}x}.$$

The P.I. solution.

Let $y = a\sin 2x + b\cos 2x$

$$\frac{dy}{dx} = 2a\cos 2x - 2b\sin 2x$$

$$\frac{d^2y}{dx^2} = -4a\sin 2x - 4b\cos 2x$$

Substituting these values in the differential equation

$$6(-4a\sin 2x - 4b\cos 2x) - (2a\cos 2x - 2b\sin 2x) - (a\sin 2x + b\cos 2x) = 10\sin 2x$$

$$-24a\sin 2x - 24b\cos 2x - 2a\cos 2x + 2b\sin 2x - a\sin 2x - b\cos 2x = 10\sin 2x.$$

Equating coefficients of $\sin 2x$ and $\cos 2x$

$$-24a + 2b - a = 10$$

$$\boxed{-25a + 2b \quad = 10} \qquad \ldots(1)$$

$$-24b - 2a - b = 0$$

$$\boxed{-2a - 25b \quad = 0} \qquad \ldots(2)$$

Solving (1) and (2)

$$(1)\times(-2) \qquad 50a - 4b \quad = -20 \qquad \ldots(3)$$

$$(2)\times 25 \qquad -50a - 625b = 0 \qquad \ldots(4)$$

$$(3) + (4) - 629b = -20 \Rightarrow b = \frac{20}{629}$$

$$a = -\frac{25}{2}b = -\frac{25}{2}\left(\frac{20}{629}\right) = -\frac{250}{629}.$$

The general solution

y = C.F. + P.I.

$$y = Ae^{\frac{1}{2}x} + Be^{-\frac{1}{3}x} - \frac{250}{629}\sin 2x + \frac{20}{629}\cos 2x.$$

---

5. (a) The square root of $3 + 4i$ is written $\sqrt{3+4i} = \pm(a + bi)$ where a, b $\in\mathbb{R}$.

Squaring up both sides

$$3 + 4i = a^2 + b^2i^2 + 2abi = (a^2 - b^2) + 2abi.$$

Equating real and imaginary terms

$$a^2 - b^2 = 3 \qquad 2ab = 4 \Rightarrow b = \tfrac{2}{a}$$

$$a^2 - \tfrac{4}{a^2} = 3 \Rightarrow a^4 - 3a^2 - 4 = 0$$

$$a^2 = \frac{3 \pm \sqrt{9+16}}{2} = \frac{3 \pm 5}{2}$$

$$a^2 = 4 \quad \text{or} \quad a^2 = -1$$

$$a = \pm 2 \qquad a = \pm i$$

but a is real (same a as in the first two lines)

if $a = 2, b = \frac{2}{2} = 1$, if $a = -2, b = \frac{2}{-2} = -1$.

$\therefore \sqrt{3 + 4i} = \pm (2 + i)$

check $3 + 4i = 4 + i^2 + 4i = 3 + 4i$.

(b) $z^3 - 1 = 0$

$(z - 1)\left(z^2 + z + 1\right) = 0$, from which $z = 1$ or

$$z^2 + z + 1 = 0 \Rightarrow z = \frac{-1 \pm \sqrt{1 - 4}}{2} = -\frac{1}{2} \pm i\frac{\sqrt{3}}{2}$$

$$z_1 = 1,\ z_2 = -\frac{1}{2} + i\frac{\sqrt{3}}{2},\ z_3 = -\frac{1}{2} - i\frac{\sqrt{3}}{2}.$$

Alternatively

$z^3 - 1 = 0 \Rightarrow z^3 = 1 \Rightarrow z = 1^{\frac{1}{3}}$.

The cube roots of unity are found:

$z = 1\,(\cos 2k\pi + i \sin 2k\pi)^{\frac{1}{3}}$

where $k = 0, 1, 2$

$\therefore z_1 = 1,\ z_2 = 1\left(\cos \frac{2\pi}{3} + i \sin \frac{2\pi}{3}\right),\ z_3 = 1\left(\cos \frac{4\pi}{3} + i \sin \frac{4\pi}{3}\right)$

$z_1 = 1,\ z_2 = -\frac{1}{2} + i\frac{\sqrt{3}}{2},\ z_3 = -\frac{1}{2} - i\frac{\sqrt{3}}{2}$.

6.

$$\begin{aligned}
f(x) &= \sin^2 x \\
f'(x) &= 2 \sin x \cos x \\
f''(x) &= 2\cos^2 x + 2\sin x(-\sin x) = 2\cos^2 x - 2\sin^2 x \\
f'''(x) &= 4\cos x(-\sin x) - 4\sin x \cos x = -8 \sin x \cos x \\
f^{(iv)}(x) &= -8\cos^2 x + 8\sin^2 x \\
f^{(v)}(x) &= -16\cos x(-\sin x) + 16 \sin x \cos x \\
f^{(v)}(x) &= 32 \sin x \cos x = 16 \sin 2x \Rightarrow f^{(vi)}(x) = 32 \cos 2x
\end{aligned}$$

$f(0) = 0, \quad f'(0) = 0, \quad f''(0) = 2, \quad f'''(0) = 0, \quad f^{(iv)}(0) = -8,$

$f^{(v)}(0) = 0, \quad f^{(vi)}(0) = 32,$

$$\therefore \sin^2 x = x^2 - \frac{1}{3}x^4 + \frac{2}{45}x^6$$

$$\cos^2 x = 1 - \sin^2 x = 1 - x^2 + \frac{1}{3}x^4 - \frac{2}{45}x^6$$

$$\cos 2x = 1 - 2\sin^2 x = 1 - 2x^2 + \frac{2}{3}x^4 - \frac{4}{45}x^6.$$

---

7. (a) $z = -3 - 4i \quad |z| = 5$

$$\arg z = 180° + 53° 8' \quad \text{since } \tan\theta = \frac{-4}{-3}.$$

$$= 233°8'$$

$$z = 5\underline{/233°}8' = 5(\cos 233°8' + i\sin 233°8')$$

(b) $z^3 = 5(\cos 233°8' + i\sin 233°8')$

$$z = 5^{\frac{1}{3}} \underline{/\frac{233°8' + 2k\pi}{3}} \qquad k = 0, \pm 1$$

$$z_1 = 5^{\frac{1}{3}} \underline{/\frac{233°8'}{3}} = 1.25 \underline{/77.7°}$$

$$z_2 = 1.25 \underline{/197.7°} \qquad z_3 = 1.25 \underline{/-42.3°}$$

(c)

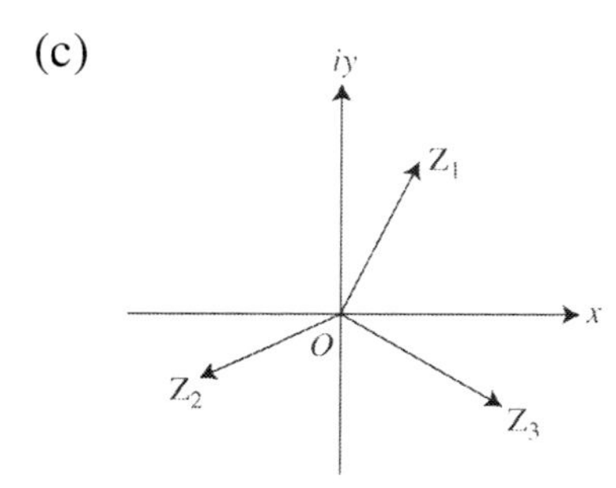

---

8. (a) $$\frac{1}{(r+3)(r+5)} \equiv \frac{A}{r+3} + \frac{B}{r+5}$$

$$1 \equiv A(r+5) + B(r+3)$$

if $r = -3$

$$1 = 2A \Rightarrow A = \tfrac{1}{2}$$

if $r = -5$

$$1 = -2B \Rightarrow B = -\tfrac{1}{2}$$

$$\therefore \frac{1}{(r+3)(r+5)} \equiv \frac{1}{2(r+3)} - \frac{1}{2(r+5)}.$$

(b) Using method of differences

when $r = 1$ $\qquad \frac{1}{2 \times 4} - \frac{1}{2 \times 6}$

$r = 2$ $\qquad \frac{1}{2 \times 5} - \frac{1}{2 \times 7}$

$r = 3$ $\qquad \frac{1}{2 \times 6} - \frac{1}{2 \times 8}$

$\dots$

$r = n-1$ $\qquad \frac{1}{2(n+2)} - \frac{1}{2(n+4)}$

$r = n$ $\qquad \frac{1}{2(n+3)} - \frac{1}{2(n+5)}$

$$\therefore \sum_{r=1}^{n} \frac{1}{(r+3)(r+5)} = \frac{1}{2 \times 4} + \frac{1}{2 \times 5} - \frac{1}{2(n+4)} - \frac{1}{2(n+5)}$$

$$= \frac{9}{40} - \frac{2n+9}{(n+4)(n+5)}$$

$$= \frac{9(n^2+9n+20) - 40(2n+9)}{40(n+4)(n+5)}$$

$$= \frac{9n^2+n-180}{40(n+4)(n+5)}.$$

Examination Test Papers.

GCE Advanced Level.

Further Pure Mathematics FP2

Test Paper 7

Time: 1 hour and 30 minutes.

Instructions and Information

Candidates may use any calculator allowed by the regulations of their Examination Board.

Full marks are awarded for correct answers to ALL questions.

This paper has eight questions.

You can start working with any question and you must label clearly all parts.

1. (a) Sketch the curves with equations

$r = \cos\theta$ and $r = \sin\theta$.

(b) Find the polar coordinates of the points of intersection of the two curves.

(c) Find the exact value of the area of the finite region bounded by the two curves. **(8)**

---

2. Expand $\cos(x + h)$ in powers of h by Taylor's Theorem.

By putting $x = \frac{\pi}{3}$ and $h = \left(1^\circ \times \frac{\pi}{180}\right)^c$, deduce the value of $\cos 61^\circ$

correct to 5 decimal places. **(10)**

---

3. (a) Sketch the locus of z and given the Cartesian equation of the locus of z when:

(i) $|z + 3| = |z + 1|$ (ii) $|z + 2| = |z + 1 - i|$ (iii) $|z| = |z - 1|$. **(12)**

---

4. Solve $\frac{d^2y}{dx^2} - 2\frac{dy}{dx} + 5y = 5e^{2x}$. **(10)**

---

5. Factorize $z^5 + 1 = 0$ in a linear and two quadratic factors and hence

find the roots of the quadratic equation. Use $\cos\theta + i\sin\theta = \angle\theta$ for short hand.

Show these roots on an Argand diagram. **(8)**

---

6. Find the general solution $\frac{dx}{d\theta} + \cot\theta x = \cos\theta$.

If $\theta = \frac{\pi}{2}$ for $x = 1$. **(8)**

---

7. (a) Express $\left(\cos\frac{3\pi}{4} + i\sin\frac{3\pi}{4}\right) \times \left(\cos\frac{\pi}{4} + i\sin\frac{\pi}{4}\right)$ in the cartesian form.

(b) Express $\frac{\cos 3\theta + i\sin 3\theta}{\cos\theta + i\sin\theta}$ in the form $x + iy$. **(10)**

---

8. Find the set of real values of x for which $|x + 4| > 2|x - 3|$. **(9)**

TOTAL FOR PAPER: 75 MARKS

Examination Test Papers.

GCE Advanced Level.

Test Paper 7 Solutions

Further Pure Mathematics FP2

1. (a)

$r = \sin\theta$

$\theta = \frac{\pi}{4}$

P

$r = \cos\theta$

O initial line

(b) $\sin\theta = \cos\theta$

$\tan\theta = 1$

$$\theta = \frac{\pi}{4} \Rightarrow r = \frac{1}{\sqrt{2}} \qquad P\left(\frac{1}{\sqrt{2}}, \frac{\pi}{4}\right), \quad O\,(0,0)$$

(c) $$2 \times \frac{1}{2}\int_0^{\frac{\pi}{4}} r^2 \mathrm{d}\theta = \int_0^{\frac{\pi}{4}} \cos^2\theta \, \mathrm{d}\theta = \frac{1}{2}\int_0^{\frac{\pi}{4}} (\cos 2\theta + 1)\, \mathrm{d}\theta$$

$\cos 2\theta = 2\cos^2\theta - 1$

$\cos^2\theta = \dfrac{\cos 2\theta + 1}{2}$

$$= \frac{1}{2}\left[\frac{\sin 2\theta}{2} + \theta\right]_0^{\frac{\pi}{4}}$$

$$= \frac{1}{2}\left[\frac{\sin\frac{\pi}{2}}{2} + \frac{\pi}{4}\right] - 0$$

$$= \frac{1}{4} + \frac{\pi}{8}$$

$$= \frac{1}{4}\left(1 + \frac{\pi}{2}\right) \text{ s.u.}$$

---

2. Let $\mathrm{f}(x+h) = \cos(x+h)$

$\mathrm{f}(x) = \cos x, \quad \mathrm{f}'(x) = -\sin x, \quad \mathrm{f}''(x) = -\cos x \quad \mathrm{f}'''(x) = \sin x$

Taylor's Theorem gives

$$\cos(x+h) = \cos x + h(-\sin x) + \frac{h^2}{2!}(-\cos x) + \frac{h^3}{3!}\sin x$$

$$\cos(x+h) = \cos x - h\sin x - \frac{h^2}{2}\cos x + \frac{h^3}{6}\sin x.$$

Let $x = 60° = \dfrac{\pi}{3}$

$$\cos(60° + h) = \cos 60° - h\sin 60° - \frac{h^2}{2}\cos 60° + \frac{h^3}{6}\sin 60°$$

$$= 0.5 - h\,0.866 - \frac{h^2}{2}0.5 + \frac{h^3}{6}0.866.$$

If now $h = 1° = 0.0174532^c$ $\qquad h^2 = 0.00030461742$ $\qquad h^3 = 0.0000053165769$

$$\cos 61° = 0.5 - 0.074532 \times 0.866 - 0.00030461742 \times 0.25 + 7.6735927 \times 10^{-7}$$

$$= 0.4848101 \approx 0.48481.$$

From the calculator $\cos 61° \approx 0.48481$ correct to 5 d.p.

---

3. (a) (i) $\quad |z + 3| = |z + 1|.$

Let $z = x + iy$

$$|x + iy + 3| = |x + iy + 1|$$

$$\sqrt{(x+3)^2 + y^2} = \sqrt{(x+1)^2 + y^2}$$

squaring both sides

$$x^2 + 6x + 9 + y^2 = x^2 + 2x + 1 + y^2$$

$$6x + 9 = 2x + 1$$

$$\boxed{x = -2}$$

(ii) $|z + 2| = |z - 1 - i|$

$$|x + iy + 2| = |x + iy - 1 - i|$$

$$\sqrt{(x+2)^2 + y^2} = \sqrt{(x-1)^2 + (y-1)^2}$$

squaring both sides

$$x^2 + 4x + 4 + y^2 = x^2 - 2x + 1 + y^2 - 2y + 1$$

$$4x + 4 = -2x + 1 - 2y + 1$$

$$6x + 2y + 2 = 0$$

$$\boxed{3x + y + 1 = 0}$$

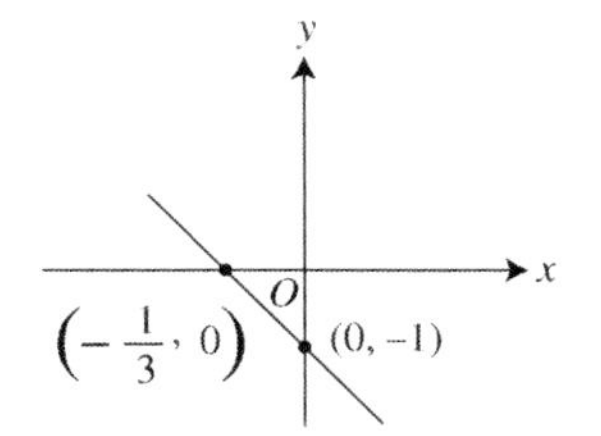

(iii) $|z| = |z-1|$

$$|x+iy| = |x+iy-1|$$

$$\sqrt{x^2+y^2} = \sqrt{(x-1)^2+y^2}$$

$$x^2+y^2 = x^2-2x+1+y^2$$

$$x = \frac{1}{2}.$$

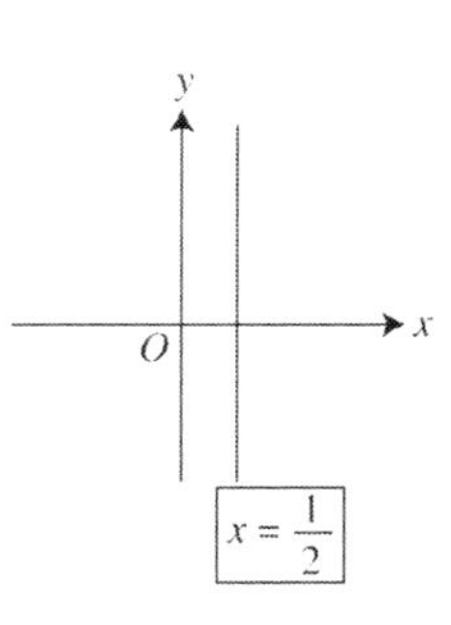

---

4. Solve $\dfrac{d^2y}{dx^2} - 2\dfrac{dy}{dx} + 5y = 5e^{2x}$.

The complementary function of $\dfrac{d^2y}{dx^2} - 2\dfrac{dy}{dx} + 5y = 0$

$m^2 - 2m + 5m = 0$

$$m = \frac{2 \pm \sqrt{4-20}}{2} = 1 \pm i2$$

$m_1 = 1 + i2$ and $m_2 = 1 - i2$

$a = 1$ and $b = 2$

the roots are complex

$y = e^x(A\cos 2x + B\sin 2x)$.

The particular integral.

let $y = Ae^{2x}$ $\quad \dfrac{dy}{dx} = 2Ae^{2x}$ $\quad \dfrac{d^2y}{dx^2} = 4Ae^{2x}$

substituting these values

$4Ae^{2x} - 2\left(2Ae^{2x}\right) + 5\left(Ae^{2x}\right) = 5e^{2x}$

$4Ae^{2x} - 4Ae^{2x} + 5A^{2x} = 5e^{2x}$.

Equate coefficients

$4A - 4A + 5A = 5 \Rightarrow A = 1$.

Therefore the P.I. solution is $y = e^{2x}$,

$\therefore y = e^x\left(A\cos 2x + B\sin 2x\right) + e^{2x}$.

---

5. $z^5 + 1 = 0 \Rightarrow z^5 = -1 = \cos\pi + i\sin\pi$

$z^5 = \underline{/\pi} \Rightarrow z = \underline{/(\pi+2k\pi)^{\frac{1}{5}}} = \underline{/\frac{\pi+2k\pi}{5}}$

where k = 0, ±1, ±2

$z_1 = \underline{/\frac{\pi}{5}}, \quad z_2 = \underline{/\frac{3\pi}{5}}, \quad z_3 = \underline{/-\frac{\pi}{5}}, \quad z_4 = \underline{/\pi}, \quad z_5 = \underline{/-\frac{3\pi}{5}}$

$$\left(z - \cos\tfrac{\pi}{5} - i\sin\tfrac{\pi}{5}\right)\left(z - \cos\tfrac{3\pi}{5} - i\sin\tfrac{3\pi}{5}\right)\left(z - \cos\tfrac{\pi}{5} + i\sin\tfrac{\pi}{5}\right)\left(z - \cos\tfrac{3\pi}{5} + i\sin\tfrac{3\pi}{5}\right)$$

$$(z - \cos\pi - i\sin\pi) = (z+1)\left(z^2 - 2z\cos\tfrac{\pi}{5} + 1\right)\left(z^2 - 2z\cos\tfrac{3\pi}{5} + 1\right) = 0$$

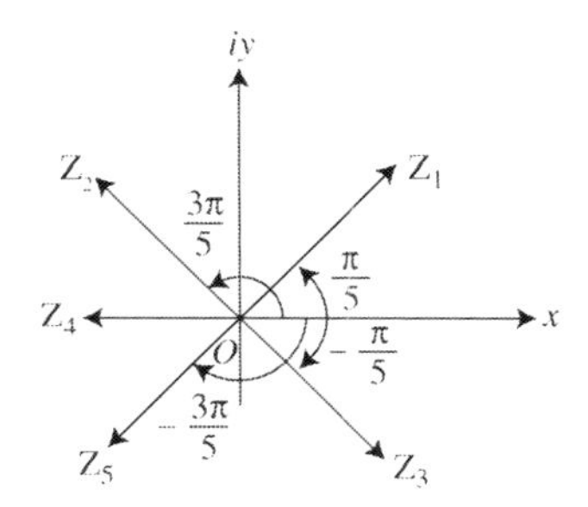

---

6. $\dfrac{dx}{d\theta} + (\cot\theta)x = \cos\theta$

$\text{I.F.} = e^{\int \cot\theta \, d\theta} = e^{\ln\sin\theta} = \sin\theta$

$$\frac{d}{d\theta}(x\sin\theta) = \frac{\sin 2\theta}{2}$$

$$\int d(x\sin\theta) = \int \frac{\sin 2\theta}{2} d\theta$$

$$x\sin\theta = \frac{\cos 2\theta}{4} + c$$

$$1.\ \sin\tfrac{\pi}{2} = \cos\frac{2\theta}{4} + c \Rightarrow c = 1 + \frac{\cos\pi}{4} = \frac{3}{4}$$

$$x = \frac{3}{4}\operatorname{cosec}\theta - \frac{1}{4}\cos 2\theta \operatorname{cosec}\theta.$$

---

7. (a)

$$\left(\cos\frac{3\pi}{4} + i\sin\frac{3\pi}{4}\right) \times \left(\cos\frac{\pi}{4} + i\sin\frac{\pi}{4}\right)$$

$$= \cos\left(\frac{3\pi}{4} + \frac{\pi}{4}\right) + i\sin\left(\frac{3\pi}{4} + \frac{\pi}{4}\right)$$

$$= \cos\pi + i\sin\pi$$

$$= -1 + i0.$$

(b)

$$\frac{\cos 3\theta + i\sin 3\theta}{\cos\theta + i\sin\theta} \times \frac{\cos\theta - i\sin\theta}{\cos\theta - i\sin\theta}$$

$$= \frac{\cos 3\theta\cos\theta + i\sin 3\theta\cos\theta - i\sin\theta\cos 3\theta - i^2\sin 3\theta\sin\theta}{\cos^2\theta + \sin^2\theta}$$

$$= (\cos 3\theta\cos\theta + \sin 3\theta\sin\theta) + i(\sin 3\theta\cos\theta - \sin\theta\cos 3\theta)$$

$$= \cos(3\theta - \theta) + i\sin(3\theta - \theta)$$

$$= \cos 2\theta + i\sin 2\theta.$$

---

8. $|x+4| > 2|x-3|$ squaring up both sides $(x+4)^2 > 4(x-3)^2$

$$(x+4)^2 - 4(x-3)^2 > 0$$

$$\left[(x+4) - 2(x-3)\right]\left[(x+4) + 2(x-3)\right] > 0$$

$$(x + 4 - 2x + 6)(x + 4 + 2x - 6) > 0$$

$$(-x + 10)(3x - 2) > 0$$

multiplying both side by -1

$$(x - 10)(3x - 2) < 0$$

$$\frac{2}{3} < x < 10$$

the set of real values of x.

Examination Test Papers.

GCE Advanced Level.

Further Pure Mathematics FP2

Test Paper 8

Time: 1 hour and 30 minutes.

Instructions and Information

Candidates may use any calculator allowed by the regulations of their Examination Board.

Full marks are awarded for correct answers to ALL questions.

This paper has eight questions.

You can start working with any question and you must label clearly all parts.

1. (a) Sketch the polar curve $r = 1 + 2\cos\theta$.

(b) Determine the area between the inner loop and the outer loop.

Take the area of the inner loop to be $\pi - \frac{3\sqrt{3}}{2}$. **(8)**

---

2. By means of the Taylor's series method, derive the solution, as a power series of ascending powers of x as far as the term x^6, of the differential equation

$$\frac{d^2y}{dx^2} - 5x\frac{dy}{dx} + 2y = 0,$$

given that $y = 2$, $\frac{dy}{dx} = 1$ when $x = 0$. **(12)**

---

3. If $|z - 3| = \sqrt{3}\,|z + 1 - i|$.

Show that the locus of z is a circle, stating its centre and its radius.

Sketch this locus. **(8)**

---

4. Solve the second order differential equations:

(i) $\frac{d^2y}{dx^2} - 3\frac{dy}{dx} + 2y = 2e^{2x}$ (ii) $\frac{d^2y}{dx^2} + 8\frac{dy}{dx} + 16y = e^{-4x}$. **(12)**

---

5. (i) Describe the locus $|z + 1 - i2| = 5$ and hence sketch this locus on an Argand diagram.

(ii) Describe the locus of $\arg\left(\frac{z - z_1}{z - z_2}\right) = \frac{\pi}{4}$ and hence sketch if $z_1 = -1 + i$ and $z_2 = -2$. **(12)**

---

6. (a) Use de Moivre's theorem to show $\sin 5\theta = \sin\theta(16\cos^4\theta - 12\cos^2\theta + 1)$.

(b) Use de Moivre's theorem to show $\cos 5\theta = 16\cos^5\theta - 20\cos^3\theta + 5\cos\theta$. **(9)**

---

7. (a) Express $2\left(\cos\frac{\pi}{12} + i\sin\frac{\pi}{12}\right) \times 3\left(\cos\frac{3\pi}{4} + i\sin\frac{3\pi}{4}\right)$ in the form $x + iy$.

(b) Express $\dfrac{5\left(\cos\frac{5\pi}{12} + i\sin\frac{5\pi}{12}\right)}{\cos\frac{\pi}{4} + i\sin\frac{\pi}{4}}$ in the form $x + iy$. **(8)**

---

8. Find the set of real values of x for which $|x - 2| < |x + 3|$. **(6)**

TOTAL FOR PAPER: 75 MARKS

Examination Test Papers.

GCE Advanced Level.

Test Paper 8 Solutions

Further Pure Mathematics FP2

1. (a)

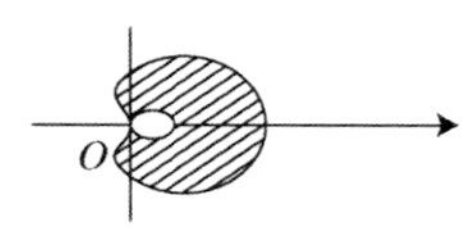

(b) $$2 \times \frac{1}{2}\int_0^{\frac{2\pi}{3}} r^2 \mathrm{d}\theta = \int_0^{\frac{2\pi}{3}} (1 + 2\cos\theta)^2 \,\mathrm{d}\theta$$

$$\cos 2\theta = 2\cos^2\theta - 1 \qquad = \int_0^{\frac{2\pi}{3}} \left(1 + 4\cos\theta + 4\cos^2\theta\right) \mathrm{d}\theta$$

$$\frac{\cos 2\theta + 1}{2} = \cos^2\theta \qquad = \int_0^{\frac{2\pi}{3}} \left[1 + 4\cos\theta + 2(\cos 2\theta + 1)\right] \mathrm{d}\theta$$

$$= \left[\theta + 4\sin\theta + \sin 2\theta + 2\theta\right]_0^{\frac{2\pi}{3}}$$

$$= 3 \times \frac{2\pi}{3} + 4\sin\frac{2\pi}{3} + \sin\frac{4\pi}{3}$$

$$= 2\pi + \frac{4\sqrt{3}}{2} - \frac{\sqrt{3}}{2}$$

$$= 2\pi + \frac{3}{2}\sqrt{3}.$$

$$\therefore \text{Area required} \quad = 2\pi + \frac{3}{2}\sqrt{3} - \left(\pi - \frac{3}{2}\sqrt{3}\right)$$

$$= \pi + 3\sqrt{3} \ \text{ s.u.}$$

---

2. $$\frac{\mathrm{d}^2y}{\mathrm{d}x^2} - 5x\frac{\mathrm{d}y}{\mathrm{d}x} + 2y = 0 \qquad \ldots(1)$$

$$\left(\frac{\mathrm{d}^2x}{\mathrm{d}^2y}\right)_0 - 5x_0\left(\frac{\mathrm{d}y}{\mathrm{d}x}\right)_0 + 2y_0 = 0$$

$$x_0 = 0, \qquad y_0 = 2, \ \left(\frac{\mathrm{d}y}{\mathrm{d}x}\right)_0 = 1$$

$$\left(\frac{\mathrm{d}^2y}{\mathrm{d}x^2}\right)_0 = -2y_0 = -2(2) = -4.$$

Differentiating (1) with respect to x

$$\frac{d^3y}{dx^3} - 5\frac{dy}{dx} - 5x\left(\frac{dy^2}{dx^2}\right) + 2\frac{dy}{dx} = 0 \qquad \ldots (2)$$

$$\left(\frac{d^3y}{dx^3}\right)_0 = 5\left(\frac{dy}{dx}\right)_0 + 5x_0\left(\frac{d^2y}{dx^2}\right)_0 - 2\left(\frac{dy}{dx}\right)_0$$
$$= 5 - 2 = 3.$$

Differentiating (2) with respect to x

$$\frac{d^4y}{dx^4} - 5\frac{d^2y}{dx^2} - 5\frac{d^2y}{dx^2} - 5x\frac{d^3y}{dx^3} + 2\frac{d^2y}{dx^2} = 0 \qquad \ldots (3)$$

$$\left(\frac{d^4y}{dx^4}\right)_0 = 10\left(\frac{d^2y}{dx^2}\right)_0 + 5x_0\left(\frac{d^3y}{dx^3}\right)_0 - 2\left(\frac{d^2y}{dx^2}\right)_0$$

$$= 10(-4) - 2(-4) = -32.$$

Differentiating (3) with respect to x

$$\frac{d^5y}{dx^5} - 10\frac{d^3y}{dx^3} - 5\frac{d^3y}{dx^3} - 5x\frac{d^4y}{dx^4} + 2\frac{d^3y}{dx^3} = 0 \qquad \ldots (4)$$

$$\frac{d^5y}{dx^5} = 15\frac{d^3y}{dx^3} + 5x\frac{d^4y}{dx^4} - 2\frac{d^3y}{dx^3}$$

$$\left(\frac{d^5y}{dx^5}\right)_0 = 15 \times 3 - 2 \times 3 = 13 \times 3.$$

Differentiating (4) with respect to x

$$\frac{d^6y}{dx^6} - 15\frac{d^4y}{dx^4} - 5\frac{d^4y}{dx^4} - 5x\frac{d^5y}{dx^5} + 2\frac{d^4y}{dx^4} = 0$$

$$\left(\frac{d^6y}{dx^6}\right)_0 = 20\left(\frac{d^4y}{dx^4}\right)_0 + 5x_0\left(\frac{d^5y}{dx^5}\right)_0 - 2\left(\frac{d^4y}{dx^4}\right)_0$$

$$= 20\,(-32) - 2\,(-32) = -32 \times 18$$

$$y \approx y_0 + \left(\frac{\mathrm{d}y}{\mathrm{d}x}\right)_0 (x - x_0) + \left(\frac{\mathrm{d}^2 y}{\mathrm{d}x^2}\right)_0 \frac{(x - x_0)^2}{2!} + ...$$

$$y = 2 + x - \frac{4x^2}{2} + 3\frac{x^3}{3!} + (-32)\frac{x^4}{4!} + 39\frac{x^5}{5!} - 32 \times \frac{18x^6}{6!}$$

$$y \approx 2 + x - 2x^2 + \frac{x^3}{2} - \frac{4}{3}x^4 + \frac{13}{40}x^5 - \frac{4}{5}x^6.$$

---

3. $|z - 3| = \sqrt{3}\,|z + 1 - i| = \sqrt{3}\,|x + iy + 1 - i|$

$$|x + iy - 3| = \sqrt{3}\sqrt{(x + 1)^2 + (y - 1)^2}$$

$$\sqrt{(x - 3)^2 + y^2} = \sqrt{3}\sqrt{(x + 1)^2 + (y - 1)^2}$$

$$x^2 - 6x + 9 + y^2 = 3\left(x^2 + 2x + 1 + y^2 - 2y + 1\right)$$

$$= 3x^2 + 6x + 3 + 3y^2 - 6y + 3$$

$$2x^2 + 2y^2 + 12x - 6y - 3 = 0$$

$$x^2 + y^2 + 6x - 3y - 1.5 = 0$$

$$(x + 3)^2 + (y - 1.5)^2 - 9 - 2.25 - 1.5 = 0$$

$$(x + 3)^2 + (y - 1.5)^2 = 12.75$$

$\mathrm{C}\left(-3, \frac{3}{2}\right)$ $\quad r = \sqrt{12.75} = 3.57$ to 3 s.f.

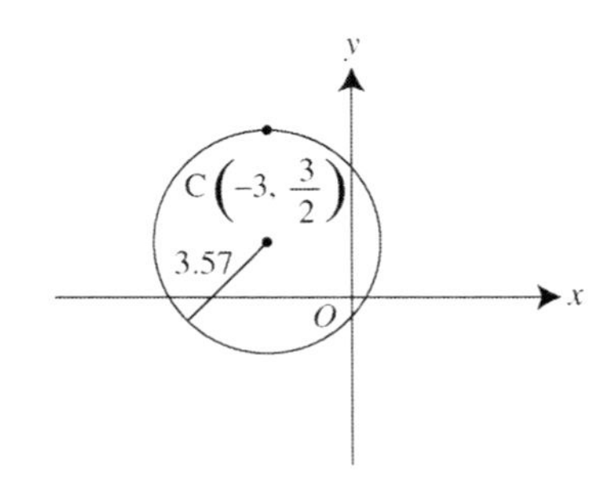

---

4. (i) $\dfrac{d^2y}{dx^2} - 3\dfrac{dy}{dx} + 2y = 2e^{2x} \qquad \ldots (1)$

$m^2 - 3m + 2 = 0$ auxiliary equation

$$m = \frac{3 \pm \sqrt{9-8}}{2} = \frac{3 \pm 1}{2} \qquad m_1 = 2, m_2 = 1$$

$y = Ae^{2x} + Be^x$ the C.F.

The particular integral

let $y = kxe^{2x} \Rightarrow \dfrac{dy}{dx} = ke^{2x} + 2kxe^{2x}$

$$\frac{d^2y}{dx^2} = 2ke^{2x} + 2ke^{2x} + 2kx2e^{2x}$$

substituting in (1)

$$4ke^{2x} + 4kxe^{2x} - 3ke^{2x} - 6kxe^{2x} + 2kxe^{2x} = 2e^{2x}$$

$ke^{2x} = 2e^{2x} \Rightarrow k = 2.$

Therefore the P.I. solution is $y = 2xe^{2x}$

and the general solution = C.F. + P.I.

$$y = Ae^{2x} + Be^x + 2xe^{2x}.$$

(ii) $\dfrac{d^2y}{dx^2} + 8\dfrac{dy}{dx} + 16y = e^{-4x} \qquad \ldots (2)$

The C.F. can be found if $\dfrac{d^2y}{dx^2} + 8\dfrac{dy}{dx} + 16y = 0,$

the auxiliary equation is $\qquad m^2 + 8m + 16 = 0.$

$(m+4)^2 = 0, \qquad m = -4$, equal roots.

The C.F. is $y = e^{-4x}(A + Bx)$.

The P.I. can be found if $\qquad y = kxe^{-4x}, \dfrac{dy}{dx} = -4kxe^{-4x} + ke^{-4x},$

$$\frac{d^2y}{dx^2} = -4ke^{-4x} + 16kxe^{-4x} - 4ke^{-4x}.$$

Substituting in (2) we have

$$-4ke^{-4x} + 16xke^{-4x} - 4ke^{-4x} - 32kxe^{-4x} + 8ke^{-4x} + 16kxe^{-4x} = e^{-4x}$$

$0 = 1$, which is inconsistent as LHS = 0 and

therefore we try $y = kx^2e^{-4x}$

$$\frac{dy}{dx} = 2kxe^{-4x} - 4kx^2e^{-4x}, \quad \frac{d^2y}{dx^2} = 2ke^{-4x} - 8kxe^{-4x} - 8kxe^{-4x} + 16kx^2e^{-4x}.$$

Substituting in (2) we have

$$2ke^{-4x} - 8kxe^{-4x} - 8kxe^{-4x} + 16kx^2e^{-4x} + 16kxe^{-4x} - 32kx^2e^{-4x} + 16kx^2e^{-4x} = e^{-4x}$$

$2k = 1, k = \frac{1}{2}$ therefore $y = \frac{1}{2}x^2e^{-4x}$ is the P.I.

The general solution is $y = e^{-4x}(A + Bx) + \frac{1}{2}x^2e^{-4x}$.

---

5. (i) Let $z = x + iy$ $|z + 1 - i2| = 5$.

$|x + iy + 1 - i2| = 5$

$|(x + 1) + i(y - 2)| = 5 \Rightarrow \sqrt{(x + 1)^2 + (y - 2)^2} = 5$

squaring up both sides $(x + 1)^2 + (y - 2)^2 = 25$.

Therefore the locus is a circle with a centre $C(-1, 2)$ and radius 5.

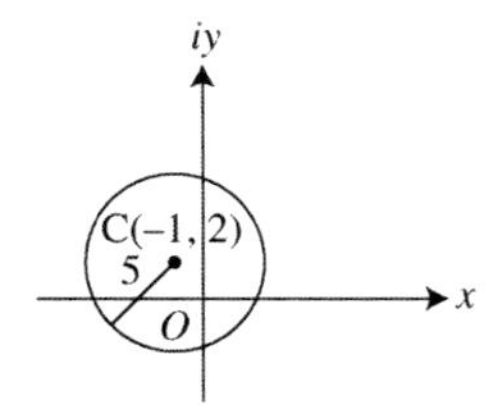

(ii) $\arg\left(\frac{z - z_1}{z - z_2}\right) = \arg(z - z_1) - \arg(z - z_2) = \frac{\pi}{4}$

if $\varphi = \arg(z - z_1)$ and $\theta = \arg(z - z_2)$ then $\varphi - \theta = \frac{\pi}{4}$,

taking tangents on both sides

$$\tan(\varphi - \theta) = \tan\frac{\pi}{4}$$

$$\frac{\tan\varphi - \tan\theta}{1 + \tan\varphi\tan\theta} = 1$$

$$\frac{m_2 - m_1}{1 + m_1 m_2} \Rightarrow \frac{\frac{y-y_1}{x-x_1} - \frac{y-y_2}{x-x_2}}{1 + \frac{y-y_1}{x-x_1}\frac{y-y_2}{x-x_2}} = 1$$

$z_1 = -1 + i$ and $z_2 = -2$

$$\arg\left(\frac{z+1-i}{z+2}\right) = \frac{\pi}{4} \text{ or } \arg(z+1-i) - \arg(z+2) = \frac{\pi}{4}$$

$$z = x + iy$$

$$\arg\left[x + 1 + i(y-1)\right] - \arg(x + 2 + iy) = \frac{\pi}{4}$$

$$\tan^{-1}\frac{y-1}{x+1} - \tan^{-1}\frac{y}{x+2} = \frac{\pi}{4}$$

taking tangents on both sides

$$\frac{\frac{y-1}{x+1} - \frac{y}{x+2}}{1 + \frac{(y-1)y}{(x+1)(x+2)}} = 1 \Rightarrow \frac{(y-1)(x+2) - y(x+1)}{(x+1)(x+2) + y(y-1)} = 1$$

$$xy - x + 2y - 2 - yx - 1 = x^2 + x + 2x + 2 + y^2 - y$$

$$x^2 + \cancel{3x} + 2 + y^2 - y - \cancel{xy} - \cancel{x} - 2y + 2 + \cancel{xy} + 1 = 0$$

$$x^2 + y^2 + 4x - 2y + 4 = 0 \Rightarrow (x+2)^2 + (y-1)^2 = 1^2$$

$(x+2)^2 + (y-1)^2 = 1^2$ is the locus which is a circle with centre

$C(-2, 1)$ and radius 1.

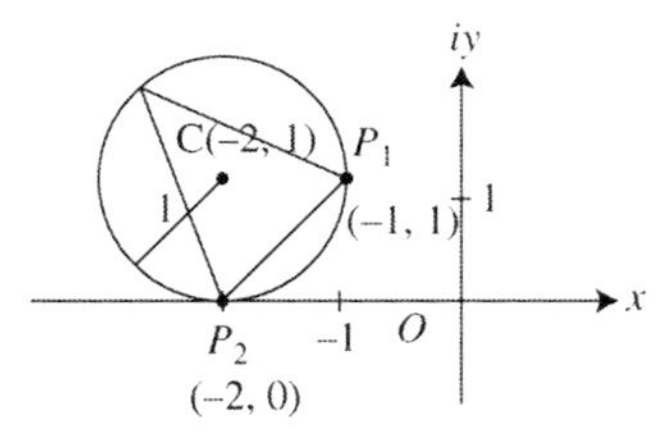

The major part of the circle is the locus.

6. $(\cos\theta + i\sin\theta)^5 = \cos 5\theta + i\sin 5\theta$

$$\text{LHS} = \cos^5\theta + 5\cos^4\theta\, i\sin\theta + \frac{5\times 4}{2}\cos^3\theta\, i^2\sin^2\theta$$

$$+\frac{5\times 4\times 3}{3!}\cos^2\theta\, i^3\sin^3\theta + \frac{5\times 4\times 3\times 2}{4!}\cos\theta\times i^4\sin^4\theta + i^5\sin^5\theta$$

$$= \cos^5\theta - 10\cos^3\theta\sin^2\theta + 5\cos\theta\sin^4\theta$$

$$+i(5\cos^4\theta\sin\theta - 10\cos^2\theta\sin^3\theta + \sin^5\theta).$$

Equating real and imaginary terms

(a) $\sin 5\theta = 5\cos^4\theta\sin\theta - 10\cos^2\theta\sin^3\theta + \sin^5\theta$

$$= \sin\theta\big[5\cos^4\theta - 10\cos^2\theta\sin^2\theta + \sin^4\theta\big]$$

$$= \sin\theta\big[5\cos^4\theta - 10\cos^2\theta(1-\cos^2\theta) + (1-\cos^2)^2\big]$$

$$= \sin\theta\big[5\cos^4\theta - 10\cos^2\theta + 10\cos^4\theta\ 1 - 2\cos^2\theta + \cos^4\theta\big]$$

$$= \sin\theta\big(16\cos^4\theta - 12\cos^2\theta + 1\big)$$

(b) $\cos 5\theta = \cos^5\theta - 10\cos^3\theta\sin^2\theta + 5\cos\theta\sin^4\theta$

$$= \cos^5\theta - 10\cos^3\theta(1-\cos^2\theta) + 5\cos\theta(1-\cos^2\theta)^2$$

$$= \cos^5\theta - 10\cos^3\theta + 10\cos^5\theta + 5\cos\theta - 10\cos^3\theta + 5\cos^5\theta$$

$$= 16\cos^5\theta - 20\cos^3\theta + 5\cos\theta.$$

---

7. (a) $2\left(\cos\frac{\pi}{12} + i\sin\frac{\pi}{12}\right)\times 3\left(\cos\frac{3\pi}{4} + i\sin\frac{3\pi}{4}\right)$

$$= 6\left(\cos\frac{\pi}{12} + i\sin\frac{\pi}{12}\right)\left(\cos\frac{9\pi}{12} + i\sin\frac{9\pi}{12}\right)$$

$$= 6\left(\cos\frac{10\pi}{12} + i\sin\frac{10\pi}{12}\right) = 6\left(\cos\frac{5\pi}{6} + i\sin\frac{5\pi}{6}\right).$$

(b) $\dfrac{5\cos\frac{5\pi}{12} + i\sin\frac{5\pi}{12}}{\cos\frac{\pi}{4} + i\sin\frac{\pi}{4}} = 5\cos\left(\dfrac{5\pi}{12} - \dfrac{3\pi}{12}\right) + i\sin\left(\dfrac{5\pi}{12} - \dfrac{3\pi}{12}\right)$

$$= 5\left(\cos\frac{\pi}{6} + i\ \sin\frac{\pi}{6}\right) = 5\left(\frac{\sqrt{3}}{2} + i\frac{1}{2}\right) = \frac{5\sqrt{3}}{2} + i\frac{5}{2}.$$

---

8. $|x - 2| < |x + 3|$.

Squaring up both sides

$$(x-2)^2 - (x+3)^2 < 0$$

$$\left[(x-2) - (x+3)\right]\left[(x-2) + (x+3)\right] < 0$$

$$-5\,(2x+1) < 0$$

$$5\,(2x+1) > 0$$

$$x > -\frac{1}{2}.$$

Examination Test Papers.

GCE Advanced Level.

Further Pure Mathematics FP2

Test Paper 9

Time: 1 hour and 30 minutes.

Instructions and Information

Candidates may use any calculator allowed by the regulations of their Examination Board.

Full marks are awarded for correct answers to ALL questions.

This paper has eight questions.

You can start working with any question and you must label clearly all parts.

1. (a) Write down the Cartesian equations of the lines:

$$\text{(i) } r = 5\sec\theta \qquad \text{(ii) } r = 4\sec\left(\theta - \tfrac{\pi}{2}\right).$$

(b) Determine the polar equation for the parabola $y^2 = 4ax$. **(8)**

---

2. (a) Find, as a series of ascending powers of x up to and including the term in x^5, an approximate solution to the differential equation $\frac{dy}{dx} = ye^{-x}$, where $y = 1$ when $x = 0$.

(b) Determine the exact solution of $\frac{dy}{dx} = ye^{-x}$, $y = 1$ where $x = 0$, by using integration.

(c) Compare the approximate and exact values of y for $x = 0, 0.5, 1, 1.5, 2$. **(12)**

---

3. (a) If $\arg(z+1) = \frac{\pi}{4}$, sketch the locus of $P(x, y)$ which is represented by x on an Argand diagram.

(b) If $\arg\left(\frac{z-2}{z-1}\right) = \frac{\pi}{3}$, find the Cartesian equation and hence sketch the locus of $P(x, y)$. **(12)**

---

4. Solve $\dfrac{d^2y}{dx^2} + \dfrac{dy}{dx} - 12y = 3x + 5$. **(10)**

---

5. Points P and Q represent the complex numbers $z = x + iy$ and $w = u + iv$ in the z-plane and the w-plane respectively .

Given that z and w are connected by the relation $w = \dfrac{z-i}{z+i}$ and that the locus of P is the x-axis, find the Cartesian equation of the locus of Q and sketch the locus of Q on an Argand diagram. **(10)**

---

6. Find $\displaystyle\sum_{r=1}^{n} \frac{1}{r(r+1)}$ using the method of differences. **(6)**

---

7. (a) Express $\dfrac{\cos 4\theta + i\sin 4\theta}{\cos 5\theta + i\sin 5\theta}$ in the form $x + iy$.

(b) $7\left(\cos\dfrac{\pi}{20} + i\sin\dfrac{\pi}{20}\right) \times 3\left(\cos\dfrac{\pi}{4} + i\sin\dfrac{\pi}{4}\right)$ in the form $x + iy$.

(you may use $\cos\theta + i\sin\theta = \angle\theta$). **(6)**

---

8. Find the set of values of x for which $|x - 2| - |x + 1| > 2$. **(11)**

TOTAL FOR PAPER: 75 MARKS

Examination Test Papers.

GCE Advanced Level.

Test Paper 9 Solutions

Further Pure Mathematics FP2

1. (i) $r = 5\sec\theta$

$$= \frac{5r}{x} = \sqrt{x^2 + y^2}$$

$$\boxed{x = 5}$$

(ii) $r = 4\sec\left(\theta - \frac{\pi}{2}\right) = \dfrac{4}{\cos\theta\cos\frac{\pi}{2} + \sin\theta\sin\frac{\pi}{2}}$

$$= \frac{4}{\sin\theta} \Rightarrow \frac{4r}{y} = r \Rightarrow \boxed{y = 4}$$

(b) $y^2 = 4ax \qquad \sin\theta = \dfrac{y}{r} \qquad \cos\theta = \dfrac{x}{r}$

$$r^2\sin^2\theta = 4ar\cos\theta$$

$$r\sin^2\theta = (4\cos\theta)\,a$$

$$r = \frac{4a\cos\theta}{\sin^2\theta} = 4a\cot\theta\,\operatorname{cosec}\theta$$

$$\boxed{r = 4a\cot\theta\,\operatorname{cosec}\theta}$$

---

2. Differentiating with respect to x the differential equation $\frac{dy}{dx} = ye^{-x}$ four times.

$$\frac{dy}{dx} = ye^{-x} \quad \ldots(1) \quad \Rightarrow \quad \frac{d^2y}{dx^2} = \frac{dy}{dx}e^{-x} - ye^{-x} \quad \ldots(2)$$

$$\frac{d^3y}{dx^3} = \frac{d^2y}{dx^2}e^{-x} - 2\frac{dy}{dx}e^{-x} + ye^{-x} \quad \ldots(3)$$

$$\frac{d^4y}{dx^4} = \frac{d^3y}{dx^3}e^{-x} - 3\frac{d^2y}{dx^2}e^{-x} + 3\frac{dy}{dx}e^{-x} - ye^{-x} \quad \ldots(4)$$

$$\frac{d^5y}{dx^5} = \frac{d^4y}{dx^4}e^{-x} - 4\frac{d^3y}{dx^3}e^{-x} + 6\frac{d^2y}{dx^2}e^{-x} - 4\frac{dy}{dx}e^{-x} + ye^{-x} \quad \ldots(5)$$

Substituting the values $y = 1$ when $x = 0$ in (1)

$\boxed{\left(\frac{\mathrm{d}y}{\mathrm{d}x}\right)_0 = 1}$ in (2) $\boxed{\left(\frac{\mathrm{d}^2y}{\mathrm{d}x^2}\right)_0 = 0}$ in (3) $\boxed{\left(\frac{\mathrm{d}^3y}{\mathrm{d}x^3}\right)_0 = -1}$

in (4) $\boxed{\left(\frac{\mathrm{d}^4y}{\mathrm{d}x^4}\right)_0 = 1}$ and in (5) $\boxed{\left(\frac{\mathrm{d}^5y}{\mathrm{d}x^5}\right)_0 = 2}$

Applying Taylor's expansion of $\mathrm{f}(x)$ about $x - x_0$

$$y \approx y_0 + \left(\frac{\mathrm{d}y}{\mathrm{d}x}\right)_0 (x - x_0) + \left(\frac{\mathrm{d}^2y}{\mathrm{d}x^2}\right)_0 \frac{(x - x_0)^2}{2!} + \left(\frac{\mathrm{d}^3y}{\mathrm{d}x^3}\right)_0 \frac{(x - x_0)^3}{3!} + \ldots$$

$$y \approx 1 + x - \frac{1}{6}x^3 + \frac{1}{24}x^4 + \frac{1}{60}x^5.$$

(b) $\frac{\mathrm{d}y}{\mathrm{d}x} = ye^{-x}$

$$\int \frac{\mathrm{d}y}{y} = \int e^{-x}\mathrm{d}x \Rightarrow \ln y = -e^{-x} + c,\ y = 1 \text{ when } x = 0$$

$\boxed{y = e^{-e^{-x}+1}}$ exact

(c) Approximate values Exact values

$y \approx 1 + x - \frac{1}{6}x^3 + \frac{1}{24}x^4 + \frac{1}{60}x^5$ $\qquad y = e^{-e^{-x}+1}$

	Approximate	Exact
$x = 0$	$y = 1$	$y = 1$
$x = 0.5$	$y \approx 1.4822917$	$y = 1.4821138$
$x = 1$	$y \approx 1.8916667$	$y = 1.8815964$
$x = 1.5$	$y \approx 2.275$	$y = 2.1746546$
$x = 2$	$y \approx 1.866667$	$y = 2.3742099.$

For values $x \leq 1$ the results are approximately correct by the two methods

for values $x > 1$ the errors are large.

---

3. (a) $\arg(z+1) = \frac{\pi}{4} \Rightarrow \arg(x+iy+1) = \frac{\pi}{4} \Rightarrow \tan^{-1}\frac{y}{x+1} = \frac{\pi}{4}$

$\tan\frac{\pi}{4} = \frac{y}{1+x} = 1 \Rightarrow \boxed{y = x+1}$ the locus

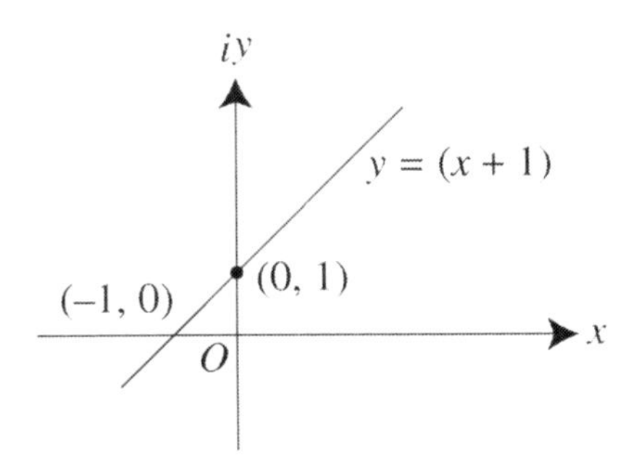

(b) $\arg\left(\frac{z-2}{z-1}\right) = \frac{\pi}{3} \Rightarrow \arg(x+iy-2) - \arg(x+iy-1) = \frac{\pi}{3}$

$\tan^{-1}\frac{y}{x-2} - \tan\frac{y}{x-1} = \frac{\pi}{3}$

taking tangents on both sides

$\frac{\frac{y}{x-2} - \frac{y}{x-1}}{1 + \frac{y}{x-2}\frac{y}{x-1}} = \tan\frac{\pi}{3} \Rightarrow \frac{y(x-1) - y(x-2)}{(x-2)(x-1)+y^2} = \sqrt{3}$

$\frac{xy - y - xy + 2y}{x^2 - 3x + 2 + y^2} = \sqrt{3} \Rightarrow \frac{y}{x^2 - 3x + y^2 + 2} = \sqrt{3}$

$y = \sqrt{3}x^2 + \sqrt{3}y^2 - 3\sqrt{3}x + 2\sqrt{3} = 0$

$x^2 + y^2 - 3x - \frac{1}{\sqrt{3}}y + 2 = 0 \Rightarrow \left(x - \frac{3}{2}\right)^2 - \frac{9}{4} + \left(y - \frac{1}{2\sqrt{3}}\right)^2 - \frac{1}{12} + 2 = 0$

$\left(x - \frac{3}{2}\right)^2 + \left(y - \frac{1}{2\sqrt{3}}\right)^2 = \frac{9}{4} + \frac{1}{12} - 2 = \frac{27+1-24}{12} = \frac{1}{3} = r^2$

$\therefore C\left(\frac{3}{2}, \frac{1}{2\sqrt{3}}\right) \quad \boxed{r = \frac{1}{\sqrt{3}}}$

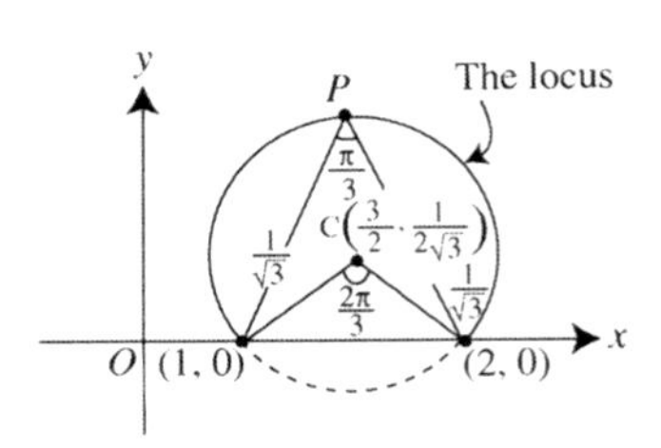

---

4. There are two steps in solving this differential equation $\frac{d^2y}{dx^2} + \frac{dy}{dx} - 12y = 3x + 5$.

(I) $\frac{d^2y}{dx^2} + \frac{dy}{dx} - 12y = 0$

the auxiliary equation $m^2 + m - 12 = 0$ which factorises to

$(m - 3)(m + 4) = 0$, $m_1 = 3$ and $m_2 = -4$.

The complementary function is $y = Ae^{3x} + Be^{-4x}$.

(II) The particular indegral can be found as follows:

Let $y = ax + b$, a function of the same form as $y = 3x + 5$.

$\frac{dy}{dx} = a, \quad \frac{d^2y}{dx^2} = 0.$

Substituting $\frac{d^2y}{dx^2} = 0$, $\frac{dy}{dx} = a$, $y = ax + b$

in the left hand side of the equation to be solved , we have

$a - 12(ax + b) = 3x + 5$

$-12ax - 12b + a = 3x + 5$

equating coefficients

$-12a = 3$ and $a = -\frac{1}{4}$

$a - 12b = 5$ and $-12b = 5 + \frac{1}{4} = \frac{21}{4}$, $b = -\frac{7}{16}$.

Therefore, the P.I. solutions is given $y = -\frac{1}{4}x - \frac{7}{16}$.

The general solution

$$y = \text{C.F.} + \text{P.I.} = Ae^{3x} + Be^{-4x} - \frac{1}{4}x - \frac{7}{16}.$$

$$\boxed{y = Ae^{3x} + Be^{-4x} - \frac{1}{4}x - \frac{7}{16}}$$

---

5. $w = \dfrac{z - i}{z + i}$ if $z = x + iy$ then

$$w = \frac{x + iy - i}{x + iy + i} = \frac{x + i(y-1)}{x + i(y+1)} \times \frac{x - i(y+1)}{x - i(y+1)}$$

$$= \frac{\left[x + i(y-1)\right]\left[x - i(y+1)\right]}{x^2 + (y+1)^2} = u + i\text{v}.$$

The locus of P is the x-axis, that is, $y = 0$

$$w = \frac{(x-i)(x-i)}{x^2+1} = \frac{x^2 + i^2 - 2ix}{x^2+1} = \frac{x^2-1}{x^2+1} - 2i\frac{x}{x^2+1} = u + i\,\text{v}.$$

Equating real and imaginary terms

$$u = \frac{x^2-1}{x^2+1} \quad \ldots(1) \quad \text{and} \quad \text{v} = \frac{2x}{x^2+1} \quad \ldots(2)$$

eliminating x from these equations by squaring up both sides of equations (1) and (2)

$$u^2 = \left(\frac{x^2-1}{x^2+1}\right)^2 \quad \ldots(3) \qquad \text{v}^2 = \frac{4x^2}{\left(x^2+1\right)^2} \quad \ldots(4)$$

adding (3) and (4) $\boxed{u^2 + \text{v}^2 = 1}$

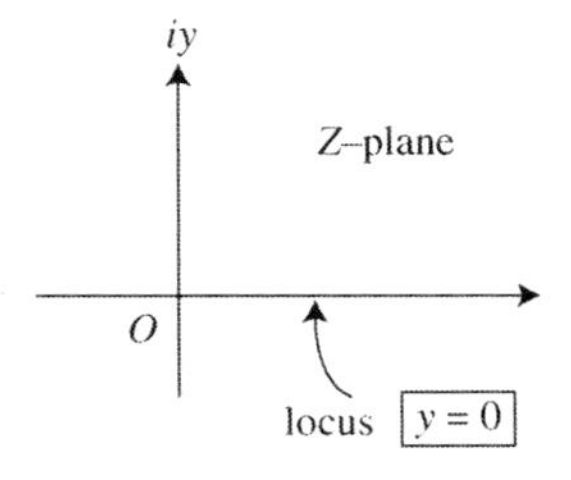

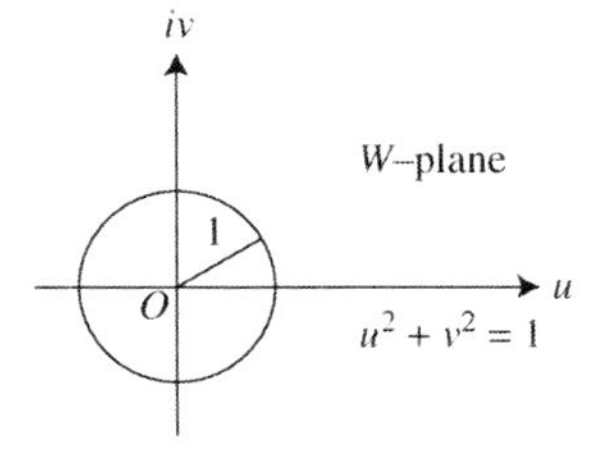

---

6. $\sum_{r=1}^{n} \frac{1}{r}(r+1)$

$$\frac{1}{r(r+1)} \equiv \frac{A}{r} + \frac{B}{r+1} \equiv \frac{A(r+1)+Br}{r(r+1)}$$

$$1 \equiv A(r+1) + Br.$$

If $r = 0$, $A = 1$, if $r = -1$, $B = -1$

$$\frac{1}{r(r+1)} \equiv \frac{1}{r} - \frac{1}{r+1}$$

$$\sum_{r=1}^{n} \frac{1}{r(r+1)} = \sum_{r=1}^{n} \left[\frac{1}{r} - \frac{1}{r+1}\right] = 1 - \frac{1}{n+1} = \frac{n+1-1}{n+1} = \frac{n}{n+1}$$

if $r = 1$ $\quad \frac{1}{1} - \frac{1}{2}$

if $r = 2$ $\quad \frac{1}{2} - \frac{1}{3}$

if $r = 3$ $\quad \frac{1}{3} - \frac{1}{4}$

$\vdots$ $\quad \vdots$

if $r = n$ $\quad \frac{1}{n} - \frac{1}{n+1}.$

---

7. (a) $\frac{\cos 4\theta + i \sin 4\theta}{\cos 5\theta + i \sin 5\theta} = \frac{\angle 4\theta}{\angle 5\theta} = \angle -\theta$

$= \cos(-\theta) + i \sin(-\theta) = \cos\theta - i \sin\theta.$

(b) $7\left(\cos\frac{\pi}{20} + i \sin\frac{\pi}{20}\right) \times 3\left(\cos\frac{\pi}{4} + i \sin\frac{\pi}{4}\right)$

$= 21\angle\frac{\pi}{20} \ \angle\frac{\pi}{4} = 21\angle\frac{\pi}{20} + \frac{5\pi}{20}$

$= 21\angle\frac{6\pi}{20} = 21\angle\frac{3\pi}{10} = 21\left(\cos\frac{3\pi}{10} + i \sin\frac{3\pi}{10}\right)$

$= 21 \cos\frac{3\pi}{10} + i21 \sin\frac{3\pi}{10} = 12.3 + i17.0$ to 3 s.f.

8.

$|x-2| - |x+1| > 2$

$|x-2| > 2 + |x+1|$

$(x-2)^2 > 4 + (x+1)^2 + 4|x+1|$

$(x-2)^2 - (x+1)^2 > 4 + 4|x+1|$

$(x-2-x-1)(x-2+x+1) > 4 + 4|x+1|$

$-3(2x-1) > 4 + 4|x+1|$

$-6x + 3 - 4 > 4|x+1|$

$-6x - 1 > 4|x+1|$

squaring up both sides

$(-6x-1)^2 > 16(x+1)^2$

$\left[(6x+1) - 4(x+1)\right][(6x+1) + 4(x+1)] > 0$

$(2x-3)(10x+5) > 0$

$5(2x-3)(2x+1) > 0$

$x > \frac{3}{2}$ or $x < -\frac{1}{2}$.

Examination Test Papers.

GCE Advanced Level.

Further Pure Mathematics FP2

Test Paper 10

Time: 1 hour and 30 minutes.

Instructions and Information

Candidates may use any calculator allowed by the regulations of their Examination Board.

Full marks are awarded for correct answers to ALL questions.

This paper has eight questions.

You can start working with any question and you must label clearly all parts.

1. By considering the stationary values of $r \sin\theta$ find the polar coordinates of the points on the curve $r = 1 + \cos\theta$ where the tangent or tangents is or are parallel to the initial line and hence find the corresponding equations of the tangent. **(10)**

---

2. Use Taylor's theorem to obtain approximate values of the following:-

(i) $\tan 45° \, 2'$ (ii) $\log_e 1.001$. **(10)**

---

3. Sketch the half – lines:

(i) $\theta = \frac{\pi}{6}$ (ii) $\theta = \frac{\pi}{2}$ (iii) $\theta = \pi$

and give the equation of the initial line.

Determine the polar equation of a line making an angle φ with the initial line, passing through a point $P(r, \theta)$. **(10)**

---

4. Solve the second order differential equations:

(i) $\frac{\mathrm{d}^2 y}{\mathrm{d}x^2} - \frac{\mathrm{d}y}{\mathrm{d}x} - 6y = 3e^{-2x}$ (ii) $\frac{\mathrm{d}^2 y}{\mathrm{d}x^2} - 2\frac{\mathrm{d}y}{\mathrm{d}x} + 5y = e^x$. **(10)**

---

5. Given that $w = \frac{z-i}{z+i}$. Find the image in the w-plane of the circle $|z| = 3$ in the z-plane. Illustrate the 2 loci in separate Argand diagrams. **(8)**

---

6. Find $\sum_{r=1}^{n} \frac{1}{r(r+1)(r+2)}$ using the method of differences.

Hint express $\frac{1}{r(r+1)(r+2)} \equiv \frac{A}{r(r+1)} + \frac{B}{(r+1)(r+2)}$. **(8)**

---

7. (a) Evaluate $\frac{\cos\frac{3\pi}{16} + i\sin\frac{3\pi}{16}}{\cos\frac{5\pi}{4} + i\sin\frac{5\pi}{4}}$ to 3 s.f.

(b) Prove that $\sin^3\theta = -\frac{1}{4}\sin 3\theta + \frac{3}{4}\sin\theta$ and $\cos^3\theta = \frac{1}{4}\cos 3\theta + \frac{3}{4}\cos\theta$

(c) Hence find the exact values of (i) $\int_0^{\frac{\pi}{4}} \sin^3\theta \, \mathrm{d}\theta$ and $\int_0^{\frac{\pi}{3}} \cos^3\theta \, \mathrm{d}\theta$. **(10)**

---

8. Find the set of real values of x for which:

(i) $\dfrac{1}{x-3} > \dfrac{1}{x+1}$ (ii) $\dfrac{x}{x+2} < 0.$ **(9)**

TOTAL FOR PAPER: 75 MARKS

Examination Test Papers.

GCE Advanced Level.

Test Paper 10 Solutions

Further Pure Mathematics FP2

1. Differentiating $r \sin\theta$ with respect to θ

$$\frac{\mathrm{d}}{\mathrm{d}\theta}(r\sin\theta) = \frac{\mathrm{d}r}{\mathrm{d}\theta}\sin\theta + r\cos\theta.$$

For stationary values $\frac{\mathrm{d}r}{\mathrm{d}\theta}\sin\theta + r\cos\theta = 0$ but $\frac{\mathrm{d}r}{\mathrm{d}\theta} = \mathrm{f}'(\theta),\ r = \mathrm{f}(\theta)$

$$\mathrm{f}'(\theta)\sin\theta + \mathrm{f}(\theta)\cos\theta \qquad \ldots(1)$$

from which we can find the coordinates of the points at which the equations of the tangents are parallel to the initial line of the curve $\quad r = 1 + \cos\theta \qquad \ldots(1)$

$\frac{\mathrm{d}r}{\mathrm{d}\theta} = -\sin\theta$ substituting this value in (1)

$$(-\sin\theta)\sin\theta + (1+\cos\theta)\cos\theta = 0 \quad -\sin^2\theta + \cos\theta + \cos^2\theta = 0$$

$$\cos\theta + \cos^2\theta - \left(1 - \cos^2\theta\right) = 0 \quad 2\cos^2\theta + \cos\theta - 1 = 0$$

$$\cos\theta = \frac{-1 \pm \sqrt{1+8}}{4} = \frac{-1 \pm 3}{4}$$

$$\cos\theta = -1, \quad \theta = \pi; \quad \cos\theta = \frac{1}{2}, \quad \theta = \frac{\pi}{3} \text{ or } -\frac{\pi}{3}.$$

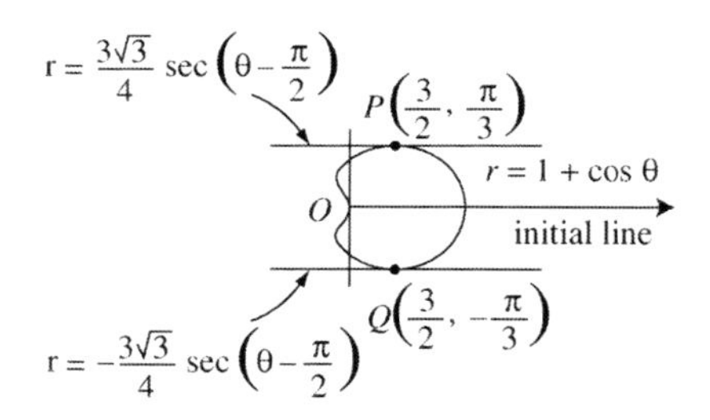

$$r = 1 + \cos\frac{\pi}{3} = 1 + \frac{1}{2} = \frac{3}{2} \qquad r = \frac{3}{2} \text{ when } \theta = \frac{\pi}{3}$$

$$r = 1 - \cos\frac{\pi}{3} = 1 + \frac{1}{2} = \frac{3}{2} \qquad r = \frac{3}{2} \text{ when } \theta = -\frac{\pi}{3}.$$

The equation of the tangent at $P\left(\frac{3}{2}, \frac{\pi}{3}\right)$.

$$r\sin\theta = \left(1+\cos\frac{\pi}{3}\right)\sin\frac{\pi}{3} \quad \text{or} \quad r\sin\theta = \frac{3\sqrt{3}}{4}$$

$$r = \frac{3\sqrt{3}}{4}\operatorname{cosec}\theta = \frac{3\sqrt{3}}{4}\sec\left(\theta - \frac{\pi}{2}\right)$$

$$\boxed{r = \frac{3\sqrt{3}}{4}\sec\left(\theta - \frac{\pi}{2}\right)}$$

The equation of the tangent at $Q\left(\frac{3}{2}, -\frac{\pi}{3}\right)$

$$r\sin\theta = \left[1+\cos\left(-\frac{\pi}{3}\right)\right]\sin\left(-\frac{\pi}{3}\right)$$

$$r\sin\theta = \left(1+\frac{1}{2}\right)\left(-\frac{\sqrt{3}}{2}\right) = -\frac{3\sqrt{3}}{4} \Rightarrow r = -\frac{3\sqrt{3}}{4}\sec\left(\theta - \frac{\pi}{2}\right).$$

---

2. Taylor's theorem states:

$$f(x+h) = f(x) + h\,\frac{f'(x)}{1!} + h^2\,\frac{f''(x)}{2!} + \ldots$$

(i) $\tan 45^\circ\, 2' = \tan\left(45^\circ + \frac{2}{60}\right)^0 = \tan\left(45^\circ + \frac{1}{30}\right)^0$

$$= \tan\left(\frac{\pi}{4} + \frac{\pi}{30\times 180}\right)^c = \tan\left(\frac{\pi}{4} + \frac{\pi}{5400}\right)^c$$

$$\tan\left(\frac{\pi}{4} + \frac{\pi}{5400}\right)^c = \tan\frac{\pi}{4} + \frac{\pi}{5400}\sec^2\frac{\pi}{4} + \left(\frac{\pi}{5400}\right)^2\frac{2}{2!}\sec^2\frac{\pi}{4}\tan\frac{\pi}{4}$$

$$+\left(\frac{\pi}{5400}\right)^3\frac{1}{3!}\left[2\sec^4\frac{\pi}{4} + 4\sec^2\frac{\pi}{4}\tan^2\frac{\pi}{4}\right] + \ldots$$

$$= 1 + \frac{\pi}{5400}(2) + \frac{\pi^2}{5400^2}\times\frac{1}{2}\times 2(2)\times 1\times\frac{\pi^3}{5400^3}\times\frac{1}{6}\times 4(2)(1)$$

$$+\cdots = 1 + 1.1635528\times 10^{-3} + 6.769276\times 10^{-7}$$

$$\tan\left(\frac{\pi}{4} + \frac{\pi}{5400}\right)^c = 1.0011642.$$

From the calculator $\tan 45^\circ 2' = 1.0011642$ which agrees with the answer.

(ii) $\log_e 1.001 = \log_e (1 + 0.001)$

$$\log_e (1 + 0.001) = \log_e 1 + 0.001 \times \frac{1}{1 + 0.001} + \frac{0.001^2}{2}\left[-\frac{1}{(1 + 0.001)^2}\right]$$

$$+\frac{0.001^3}{3!} \times \frac{2}{(1 + 0.001)^3} + \ldots$$

$$= 9.99 \times 10^{-4} - 4.990015 \times 10^{-7} + 3.3233533 \times 10^{-10} = 9.985 \times 10^{-4}.$$

Form the calculator $\ln 1.001 = 9.9950033 \times 10^{-4}$,

which agrees to five significant figures.

3.

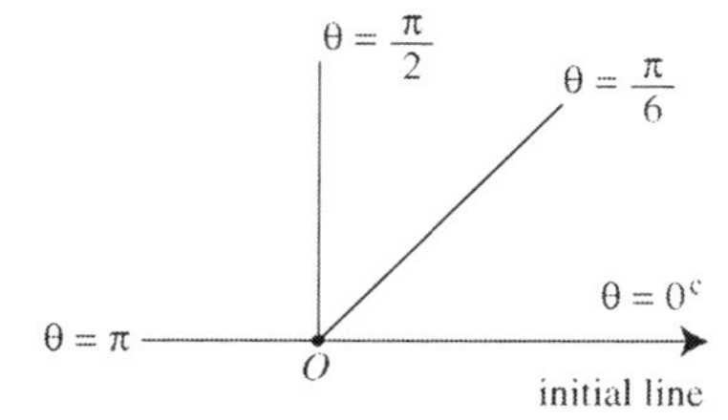

All these half-lines are passing through the pole 0.

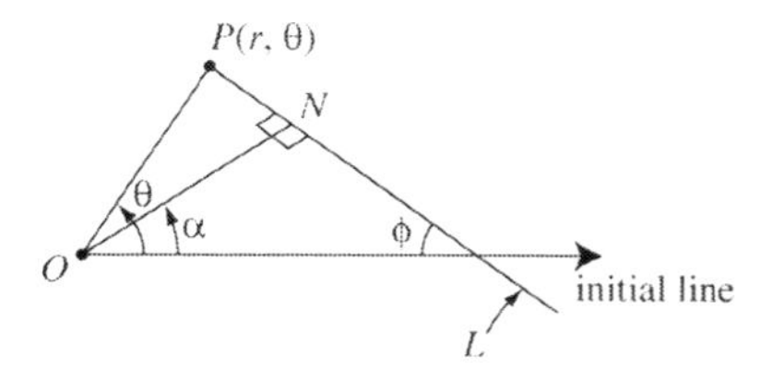

Draw a perpendicular line from the pole, 0 to the line L, making an angle α with the initial line.

Let $ON = \text{d}$, $N\hat{O}P = \theta - \alpha$

ΔONP, $\cos(\theta - \alpha) = \dfrac{ON}{OP}$

$$OP = \frac{ON}{\cos(\theta - \alpha)} = ON \sec(\theta - \alpha) = \text{d} \sec(\theta - \alpha) = r$$

$$r = \text{d} \sec(\theta - \alpha)$$

$$r = \text{d} \sec\left(\theta - \frac{\pi}{2} + \phi\right)$$

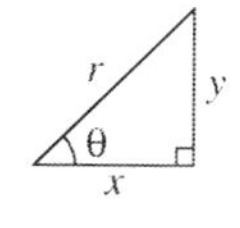

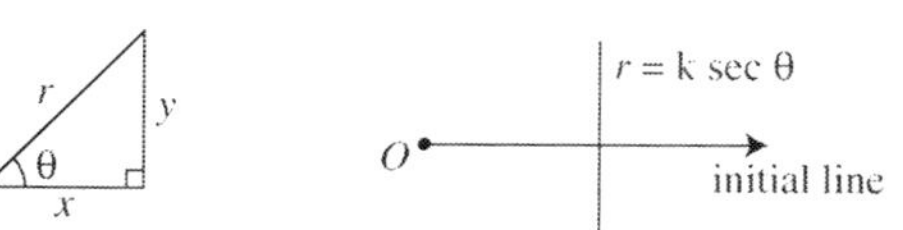

---

4. (i) $\dfrac{d^2y}{dx^2} - \dfrac{dy}{dx} - 6y = 3e^{-2x}$ (1)

The C.F. can be found if $\dfrac{d^2y}{dx^2} - \dfrac{dy}{dx} - 6y = 0$, its auxiliary equation is $m^2 - m - 6 = 0$,

$m = \dfrac{1 \pm \sqrt{1+24}}{2} = \dfrac{1 \pm 5}{2}$ $m_1 = 3$ or $m_2 = -2$.

The C.F. is $y = Ae^{3x} + Be^{-2x}$.

The P.I. can be found if $y = kxe^{-2x}$,

$\dfrac{dy}{dx} = ke^{-2x} - 2kxe^{-2x}$, $\dfrac{d^2y}{dx^2} = -2ke^{-2x} - 2ke^{-2x} + 4kxe^{-2x}$.

Substituting in (1)

$-2ke^{-2x} - 2ke^{-2x} + 4kxe^{-2x} - ke^{-2x} + 2kxe^{-2x} - 6kxe^{-2x} = 3e^{-2x}$

$-5ke^{-2x} = 3e^{-2x}$, $k = -\dfrac{3}{5}$, therefore the P.I. $y = -\dfrac{3}{5}xe^{-2x}$ and the general solution is $y = Ae^{3x} + Be^{-2x} - \dfrac{3}{5}xe^{-2x}$.

(ii) $\dfrac{d^2y}{dx^2} - 2\dfrac{dy}{dx} + 5y = e^x$ ... (2)

The C.F. can be found if $\dfrac{d^2y}{dx^2} - 2\dfrac{dy}{dx} + 5y = 0$ its auxiliary equation is

$m^2 - 2m + 5 = 0$, $m = 1 \pm i2$ and the C.F. is $y = e^x (A\cos 2x + B\sin 2x)$.

The particular integral (P.I.) is $y = ke^x$, $\dfrac{dy}{dx} = ke^x$, $\dfrac{d^2y}{dx^2} = ke^x$

$ke^x - 2ke^x + 5ke^x = e^x \Rightarrow 4k = 1$ $k = \frac{1}{4}$.

The general solution

$$\boxed{y = e^x (A\cos 2x + B\sin 2x) + \frac{1}{4}e^x.}$$

---

5. $w = \dfrac{z-i}{z+i}$ where $z = x + iy$

$$w = \frac{x+iy-i}{x+iy+i} = \frac{x+i(y-1)}{x+i(y+1)} \times \frac{x-i(y+1)}{x-i(y+1)}$$

$$= \frac{[x+i(y-1)][x-i(y+1)]}{x^2+(y+1)^2} = \frac{x^2+i(y-1)x-i(y+1)x+(y^2-1)}{x^2+y^2+2y+1}$$

$|z| = 3$

$x^2 + y^2 = 3^2$

$$w = \frac{x^2+y^2-1+i(yx-x-xy-x)}{x^2+y^2+2y+1} = \frac{9-1-2ix}{9+2y+1} = \frac{8-2ix}{10+2y}$$

$$= \frac{4-ix}{5+y} = \frac{4}{5+y} - i\frac{x}{5+y} = u + i\text{v}$$

$$u = \frac{4}{5+y} \quad \ldots(1) \qquad \text{v} = -\frac{x}{5+y} \quad \ldots(2)$$

From (1) $5 + y = \dfrac{4}{u} = -\dfrac{x}{\text{v}} \Rightarrow \boxed{x = -\dfrac{4\text{v}}{u} \quad \ldots(3)} \qquad \boxed{y = \dfrac{4}{u} - 5} \quad \ldots(4)$

squaring both sides of (3) and (4) and adding $x^2 + y^2 = \left(-\dfrac{4\text{v}}{u}\right)^2 + \left(\dfrac{4}{u} - 5\right)^2 = 9$

$$\frac{16\text{v}^2}{u^2} + \frac{16}{u^2} + 25 - \frac{40}{u} = 9 \Rightarrow 16\text{v}^2 + 16 + 25u^2 - 40u = 9u^2$$

$$\boxed{\therefore \text{v}^2 + \left(u - \frac{5}{4}\right)^2 = \left(\frac{3}{4}\right)^2}$$

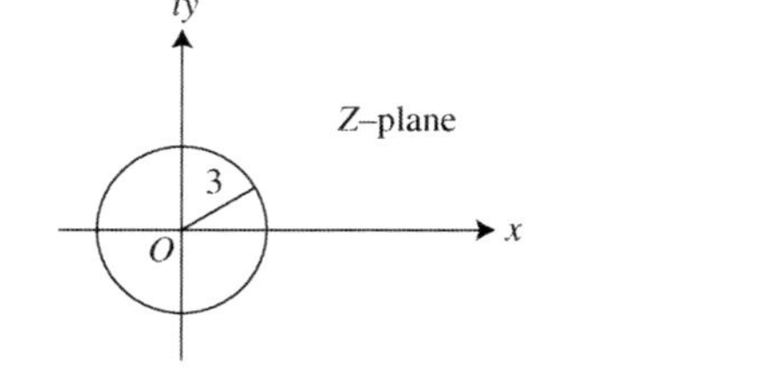

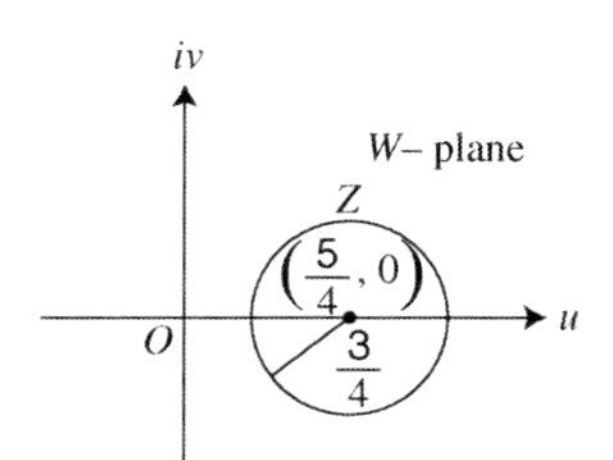

6. $$\frac{1}{r(r+1)(r+2)} \equiv \frac{A}{r(r+1)} + \frac{B}{(r+1)(r+2)} = \frac{A(r+2)}{r(r+1)(r+2)} + \frac{Br}{r(r+1)(r+2)}$$

$$1 \equiv A(r+2) + Br$$

if $r = -2$ $\quad B = -\frac{1}{2}$

if $r = 0$ $\quad A = \frac{1}{2}$

$$\therefore \frac{1}{r(r+1)(r+2)} \equiv \frac{1}{2r(r+1)} - \frac{1}{2(r+1)(r+2)}$$

$$\sum_{r=1}^{n} \frac{1}{r(r+1)(r+2)} = \sum_{r=1}^{n} \left[\frac{1}{2r(r+1)} - \frac{1}{2(r+1)(r+2)} \right]$$

for $r = 1$ $\quad \frac{1}{2 \times 1 \times 2} - \frac{1}{2 \times 2 \times 3}$

for $r = 2$ $\quad \frac{1}{2 \times 2 \times 3} - \frac{1}{2 \times 3 \times 4}$

for $r = 3$ $\quad \frac{1}{2 \times 3 \times 4} - \frac{1}{2 \times 4 \times 5}$

$\vdots \qquad \vdots$

for $r = n$ $\quad \frac{1}{2 \times n(n+1)} - \frac{1}{2 \times (n+1)(n+2)}$

$$\therefore \sum_{r=1}^{n} \frac{1}{r(r+1)(r+2)} = \frac{1}{2 \times 1 \times 2} - \frac{1}{2(n+1)(n+2)}$$

$$= \frac{(n+1)(n+2) - 2}{4(n+1)(n+2)} = \frac{n^2 + 3n + 2 - 2}{4(n+1)(n+2)} = \frac{n(n+3)}{4(n+1)(n+2)}.$$

---

7. (a) $$\frac{\cos\frac{3\pi}{16} + i\sin\frac{3\pi}{16}}{\cos\frac{5\pi}{4} + i\sin\frac{5\pi}{4}} = \frac{\angle\frac{3\pi}{16}}{\angle\frac{5\pi}{4}} = \frac{\angle\frac{3\pi}{16}}{\angle\frac{20\pi}{16}}$$

$$= \angle\left(\frac{3\pi}{16} - \frac{20\pi}{16}\right) = \angle\left(-\frac{17\pi}{16}\right) = \cos\frac{17\pi}{16} - i\sin\frac{17\pi}{16}$$

$= -0.981 - i\,(-0.195) = -0.981 + i\,0.195$ to 3 s.f.

(b) $\left(z+\frac{1}{z}\right)^3=(2\cos\theta)^3=8\cos^3\theta$

$$=z^3+3z^2\frac{1}{z}+3z\frac{1}{z^2}+\frac{1}{z^3}=z^3+\tfrac{1}{z^3}+3\left(z+\tfrac{1}{z}\right)$$

$$=2\cos 3\theta+3\,(2\cos\theta)=2\cos 3\theta+6\cos\theta$$

$$\cos^3\theta=\frac{1}{8}\,(2\cos 3\theta+6\cos\theta)=\frac{1}{4}\cos 3\theta+\frac{3}{4}\cos\theta$$

$$\left(z-\frac{1}{z}\right)^3=(2i\sin\theta)^3=8i^3\sin^3\theta=-8i\sin^3\theta=z^3-3z^2\left(\frac{1}{z}\right)+3z\left(\frac{1}{z^2}\right)-\frac{1}{z^3}$$

$$=\left(z^3-\frac{1}{z^3}\right)+3\left(-z+\frac{1}{z}\right)=\left(z^3-\frac{1}{z^3}\right)-3\left(z-\frac{1}{z}\right)$$

$$=2i\sin 3\theta-3\,(2i\sin\theta)$$

$$-8i\sin^3\theta=2i\sin 3\theta-6i\sin\theta$$

$$\sin^3\theta=-\frac{1}{4}\sin 3\theta+\frac{3}{4}\sin\theta.$$

(c) (i) $\displaystyle\int_0^{\frac{\pi}{4}}\sin^3\theta\ \mathrm{d}\theta=\int_0^{\frac{\pi}{4}}\left(-\frac{1}{4}\sin 3\theta+\frac{3}{4}\sin\theta\right)\mathrm{d}\theta$

$$=\left[\frac{\cos 3\theta}{12}-\frac{3\cos\theta}{4}\right]_0^{\frac{\pi}{4}}=\frac{\cos\frac{3\pi}{4}}{12}-\frac{3}{4}\cos\frac{\pi}{4}-\frac{\cos 0}{12}+\frac{3}{4}\cos 0$$

$$=-\frac{1}{12\sqrt{2}}-\frac{3}{4\sqrt{2}}-\frac{1}{12}+\frac{3}{4}=-\frac{10}{12\sqrt{2}}+\frac{8}{12}=-\frac{5}{6\sqrt{2}}\frac{\sqrt{2}}{\sqrt{2}}+\frac{2}{3}$$

$$=\frac{2}{3}-\frac{5\sqrt{2}}{12}=\frac{1}{12}\left(8-5\sqrt{2}\right).$$

(ii) $\displaystyle\int_0^{\frac{\pi}{3}}\left(\frac{1}{4}\cos 3\theta+\frac{3}{4}\cos\theta\right)\mathrm{d}\theta$

$$=\left[\frac{\sin 3\theta}{12}+\frac{3}{4}\sin\theta\right]_0^{\frac{\pi}{3}}=\frac{\sin\pi}{12}+\frac{3}{4}\sin\frac{\pi}{3}=\frac{3}{4}\frac{\sqrt{3}}{2}=\frac{3\sqrt{3}}{8}.$$

8. (i) $\dfrac{1}{x-3} > \dfrac{1}{x+1}$ $\quad \ldots (1)$

multiplying both sides of (1) by $(x-3)^2(x+1)^2$,

the sign of the inequality does not change

$$\frac{(x-3)^2(x+1)^2}{x-3} > \frac{(x-3)^2(x+1)^2}{x+1}$$

$$(x-3)(x+1)^2 > (x-3)^2(x+1)$$

$$(x-3)(x+1)^2 - (x-3)^2(x+1) > 0$$

$$(x-3)(x+1)\big[(x+1)-(x-3)\big] > 0$$

$$(x-3)(x+1)(x+1-x+3) > 0$$

$$4(x-3)(x+1) > 0$$

$$x > 3 \text{ or } x < -1.$$

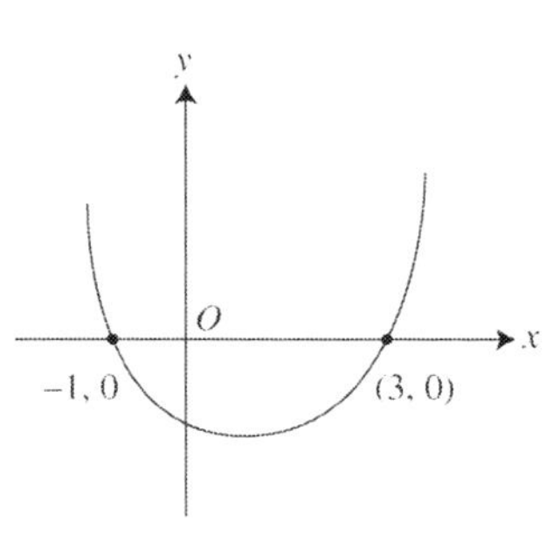

(ii) $\dfrac{x}{x+2} < 0$

multiplying both sides by $(x+2)^2$, $\dfrac{x(x+2)^2}{x+2} < 0$

$\therefore x(x+2) < 0$ and $-2 < x < 0$.

Examination Test Papers.

GCE Advanced Level.

Further Pure Mathematics FP3

Test Paper 1

Time: 1 hour and 30 minutes.

Instructions and Information

Candidates may use any calculator allowed by the regulations of their Examination Board.

Full marks are awarded for correct answers to ALL questions.

This paper has eight questions.

You can start working with any question and you must label clearly all parts.

1. Define $\sinh x$ and $\cosh x$ hence show that $\cosh^2 x - \sinh^2 x \equiv 1$. **(4)**

Prove the following identities:

(i) $\cosh(x + y) \equiv \cosh x \cosh y + \sinh x \sinh y$

(ii) $\sinh 3x \equiv 3 \sinh x + 4 \sinh^3 x$

(iii) $\cosh 3x \equiv 4 \cosh^3 x - 3 \cosh x$. **(6)**

---

2. The equation of an ellipse is given by

$$\frac{x^2}{a^2} + \frac{y^2}{b^2} = 1 \qquad \text{a > b.}$$

Sketch this curve and determine the equation of the tangent at $P(x_1, y_1)$. **(10)**

---

3. Determine the reduction formula

$nI_n = \cosh x \sinh^{n-1} x - (n - 1) I_{n-2}$.

For $I_n = \int \sinh^n x \, dx$. **(6)**

---

4. Sketch the graphs $y = 3\cosh x$ and $y = \cosh 2x$, on the same diagram.

Determine the area enclosed by these graphs. **(10)**

---

5. Solve the simultaneous linear equations using matrices.

$$x + 2y + 3z = 5$$

$$x - y - z = 0$$

$$4x + 5y + 6z = 11.$$ **(12)**

---

6. A transformation T is given by $\mathbf{y} = \mathbf{Ax}$ where $\mathbf{y}$ and $\mathbf{x}$ are column vectors

with three elements, $\mathbf{A} = \begin{pmatrix} 1 & 1 & 1 \\ 1 & 3 & 6 \\ 1 & 2 & \lambda \end{pmatrix}$, and λ is a constant.

(a) Given that $\mathbf{y} = \begin{pmatrix} 1 \\ 2 \\ 1 \end{pmatrix}$ and $\lambda = 3$, find $\mathbf{x}$.

(b) Determine the value of λ for which it is not always possible to find an $\mathbf{x}$ for a given $\mathbf{y}$. Show that in this case all vectors transform to a vector of the form $\begin{pmatrix} 2a \\ 2b \\ a+b \end{pmatrix}$. **(12)**

---

7. If $\mathbf{a} = 2\mathbf{i} - 3\mathbf{j} + 5\mathbf{k}$ $\quad \mathbf{b} = -3\mathbf{i} - 2\mathbf{k}$ $\quad \mathbf{c} = \mathbf{i} + \mathbf{j} + \mathbf{k}$.

Find

(i) $(\mathbf{a} \times \mathbf{b}) \times \mathbf{c}$

(ii) $\mathbf{a}.\,\mathbf{b}.\,\mathbf{c}$

(iii) $\mathbf{a}.\,(\mathbf{b} \times \mathbf{c})$. **(7)**

---

8. (a) Show that $\sinh x = \dfrac{2 \tanh \frac{x}{2}}{1 - \tanh^2 \frac{x}{2}} = \dfrac{2t}{1 - t^2}$.

(b) For trigonometric functions $\sin x = \dfrac{2t}{1 + t^2}$ find similar expressions for $\cos x$, $\tan x$, $\sec x$, $\operatorname{cosec} x$ and $\cot x$ using the triangle in Fig. 1

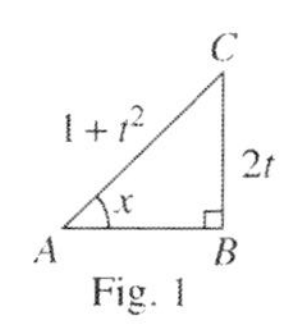

Fig. 1

Hence write down the corresponding t formulae for $\cosh x$, $\tanh x$, $\operatorname{sech} x$, $\coth x$ and $\operatorname{cosech} x$ using Osborne's rule that is replace $\sin x$ by $i \sinh x$. **(8)**

TOTAL FOR PAPER: 75 MARKS

Examination Test Papers.

GCE Advanced Level.

Test Paper 1 Solutions

Further Pure Mathematics FP3

1. $\sinh x = \dfrac{e^x - e^{-x}}{2}$ and $\cosh x = \dfrac{e^x + e^{-x}}{2}$

$$\cosh^2 x - \sinh^2 x = \left(\frac{e^x + e^{-x}}{2}\right)^2 - \left(\frac{e^x - e^{-x}}{2}\right)^2$$

$$= \frac{e^{2x} + 2 + e^{-2x} - \left(e^{2x} - 2 + e^{-2x}\right)}{2^2}$$

$$= \frac{e^{2x} + 2 + e^{-2x} - e^{2x} + 2 - e^{-2x}}{2^2}$$

$$\boxed{\cosh^2 x - \sinh^2 x \equiv 1}$$

(i) $\cosh(x + y) = \dfrac{1}{2}\left(e^{x+y} + e^{-(x+y)}\right)$

$$\cosh x \cosh y + \sinh x \sinh y = \frac{1}{2}\left(e^x + e^{-x}\right)\frac{1}{2}\left(e^y + e^{-y}\right)$$

$$+\frac{1}{2}\left(e^x - e^{-x}\right)\frac{1}{2}\left(e^y - e^{-y}\right) = \frac{1}{4}\left(e^{x+y} + e^{-x+y} + e^{x-y} + e^{-x-y}\right)$$

$$+\frac{1}{4}\left(e^{x+y} - e^{-x+y} - e^{x-y} + e^{-x-y}\right) = \frac{1}{4}\left(2e^{x+y} + 2e^{-(x+y)}\right)$$

$$= \frac{1}{2}\left(e^{x+y} + e^{-(x+y)}\right)$$

$\therefore \cosh(x + y) = \cosh x \cosh y + \sinh x \sinh y$

(ii) $\sinh 3x = 3\sinh x + 4\sinh^3 x = \dfrac{1}{2}\left(e^{3x} - e^{-3x}\right)$

$$3\sinh x + 4\sinh^3 x = \frac{3}{2}\left(e^x - e^{-x}\right) + \frac{4}{8}\left(e^x - e^{-x}\right)^3$$

$$= \frac{3}{2}e^x - \frac{3}{2}e^{-x} + \frac{1}{2}e^{3x} - \frac{3}{2}e^x + \frac{3}{2}e^{-x} - \frac{1}{2}e^{-3x}$$

$$= \frac{1}{2}\left(e^{3x} - e^{-3x}\right)$$

$\therefore \sinh 3x = 3\sinh x + 4\sin h^3 x.$

(iii) $\cosh 3x = 4\cosh^3 x - 3\cosh x = \frac{4}{8}\left(e^x + e^{-x}\right)^3 - \frac{3}{2}\left(e^x + e^{-x}\right)$

$$= \frac{1}{2}\left(e^{3x} + 3e^x + 3e^{-x} + e^{-3x}\right) - \frac{3}{2}e^x - \frac{3}{2}e^{-x}$$

$$= \frac{1}{2}(e^{3x} + e^{-3x})$$

$$\cosh 3x = \frac{1}{2}\left(e^{3x} + e^{-3x}\right)$$

$\therefore \cosh 3x = 4\cos h^3 x - 3\cosh x.$

2.

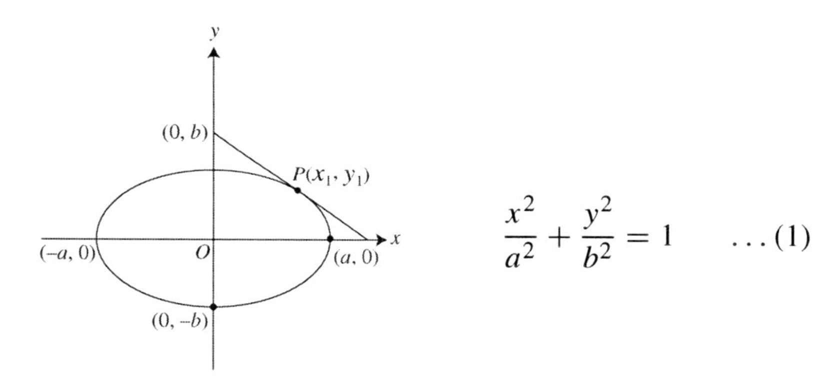

$$\frac{x^2}{a^2} + \frac{y^2}{b^2} = 1 \qquad \ldots(1)$$

Differentiating (1) with respect to x

$\frac{2x}{a^2} + \frac{2y}{b^2}\frac{dy}{dx} = 0 \Rightarrow \frac{dy}{dx} = -\frac{b^2}{a^2}\frac{x}{y}$, the gradient at P is $\frac{dy}{dx} = -\frac{b^2}{a^2}\frac{x_1}{y_1}$.

Let $y = mx + c$ be an equation of a straight line

$$y = -\frac{b^2}{a^2}\frac{x_1}{y_1}x + c$$

$$y_1 = -\frac{b^2}{a^2}\frac{x_1}{y_1}x_1 + c \Rightarrow c = \frac{{y_1}^2a^2 + b^2{x_1}^2}{a^2y_1}$$

$$\therefore y = -\frac{b^2}{a^2}\frac{x_1}{y_1}x + \frac{{y_1}^2a^2 + b^2{x_1}^2}{a^2y_1}$$

$$ya^2y_1 + b^2x_1x = {y_1}^2a^2 + b^2{x_1}^2$$

dividing each term by $a^2 b^2$

$$\frac{yy_1a^2}{a^2b^2} + \frac{x_1xb^2}{a^2b^2} = \frac{y_1a^2}{a^2b^2} + \frac{b^2x_1^{\ 2}}{a^2b^2}$$

$$\frac{yy_1}{b^2} + \frac{xx_1}{a^2} = \frac{y_1}{b^2} + \frac{x_1^{\ 2}}{a^2} = 1$$

$$\therefore \boxed{\frac{xx_1}{a^2} + \frac{yy_1}{b^2} = 1}$$

---

3. $I_n = \int \sinh^n x \, dx = \int \underset{②}{\sinh^{n-1} x} \ \underset{①}{\sinh x} \, dx = \cosh x \sinh^{n-1} x$

$$- \int \cosh x \, (n-1) \sinh^{n-2} x \cosh x \, dx$$

$$I_n = \cosh x \sinh^{n-1} x - \int \cosh^2 x \, (n-1) \sinh^{n-2} x \, dx$$

$$= \cosh x \sinh^{n-1} x - (n-1) \int \left(1 + \sinh^2 x\right) \sinh^{n-2} x \, dx$$

$$= \cosh x \sinh^{n-1} x - (n-1) \int \sinh^{n-2} x \, dx - (n-1) \int \sinh^n x \, dx$$

$$I_n (1 + n - 1) = \cosh x \sinh^{n-1} x - (n-1) I_{n-2}$$

$$\boxed{nI_n = \cosh x \sinh^{n-1} x - (n-1) I_{n-2}}$$

4.

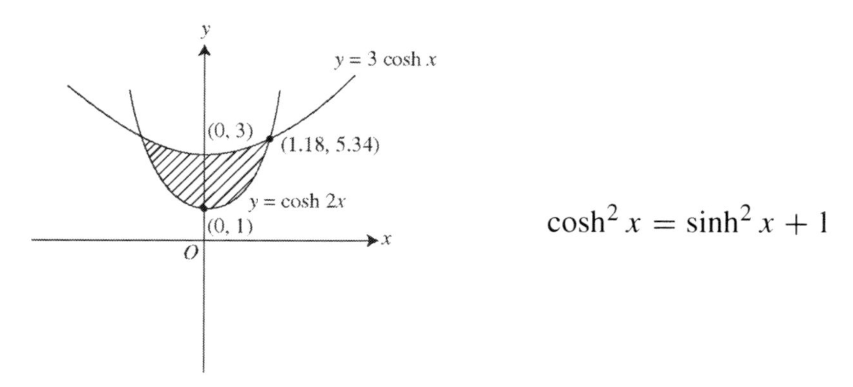

$$\cosh^2 x = \sinh^2 x + 1$$

$$3 \cosh x = \cosh 2x = \cosh^2 x + \sinh^2 x = 2 \cosh^2 x - 1$$
$$= 2 \cosh^2 x - 1$$

$$2\cosh^2 x - 3\cosh x - 1 = 0$$

$$\cosh x = \frac{3 \pm \sqrt{9+8}}{2 \times 2} = \frac{3+\sqrt{17}}{4} \text{ the negative is not defined}$$

$$\cosh x = 1.720776406 \Rightarrow x = 1.179966024$$

$$x = 1.18 \text{ to 3 s.f}$$

$$2\int_0^{1.18} 3\cosh x \, dx - 2\int_0^{1.18} \cosh 2x \, dx$$

$$= 2\left[3\sinh x - \frac{\sinh 2x}{2}\right]_0^{1.18}$$

$$= 2\,(4.420643196 - 2.624132807) = 2 \times 1.796510389 = 1.80 \times 2 \text{ to 3 s.f.}$$

$$= 3.60 \text{ to 3 s .f.}$$

5.

$$x + 2y + 3z = 5$$
$$x - y - z = 0$$
$$4x + 5y + 6z = 11.$$

The equations can be written in the matrix form

$$\begin{pmatrix} 1 & 2 & 3 \\ 1 & -1 & -1 \\ 4 & 5 & 6 \end{pmatrix} \begin{pmatrix} x \\ y \\ z \end{pmatrix} = \begin{pmatrix} 5 \\ 0 \\ 11 \end{pmatrix} \quad \ldots (1)$$

Let $\mathbf{M} = \begin{pmatrix} 1 & 2 & 3 \\ 1 & -1 & -1 \\ 4 & 5 & 6 \end{pmatrix}$.

The minors of the elements are given correspondingly:

$$\begin{vmatrix} -1 & -1 \\ 5 & 6 \end{vmatrix} = -1, \begin{vmatrix} 1 & -1 \\ 4 & 6 \end{vmatrix} = 10, \begin{vmatrix} 1 & -1 \\ 4 & 5 \end{vmatrix} = 9$$

$$\begin{vmatrix} 2 & 3 \\ 5 & 6 \end{vmatrix} = -3, \begin{vmatrix} 1 & 3 \\ 4 & 6 \end{vmatrix} = -6, \begin{vmatrix} 1 & 2 \\ 4 & 5 \end{vmatrix} = -3$$

$$\begin{vmatrix} 2 & 3 \\ -1 & -1 \end{vmatrix} = 1, \begin{vmatrix} 1 & 3 \\ 1 & -1 \end{vmatrix} = -4, \begin{vmatrix} 1 & 2 \\ 1 & -1 \end{vmatrix} = -3.$$

The minors are $\begin{pmatrix} -1 & 10 & 9 \\ -3 & -6 & -3 \\ 1 & -4 & -3 \end{pmatrix}$.

The cofactor of $\mathbf{M} = \begin{pmatrix} -1 & -10 & 9 \\ 3 & -6 & 3 \\ 1 & 4 & -3 \end{pmatrix} = \mathbf{M}^*$.

The adjoint matrix = $\mathbf{M}^{*T} = \begin{pmatrix} -1 & 3 & 1 \\ -10 & -6 & 4 \\ 9 & 3 & -3 \end{pmatrix}$

$$|\mathbf{M}| = \begin{vmatrix} 1 & 2 & 3 \\ 1 & -1 & -1 \\ 4 & 5 & 6 \end{vmatrix} = \begin{vmatrix} -1 & -1 \\ 5 & 6 \end{vmatrix} - 2\begin{vmatrix} 1 & -1 \\ 4 & 6 \end{vmatrix} + 3\begin{vmatrix} 1 & -1 \\ 4 & 5 \end{vmatrix}$$

$$= (-6+5) - 2(6+4) + 3(5+4) = -1 - 20 + 27 = 6$$

$$\mathbf{M}^{-1} = \frac{\mathbf{M}^{*T}}{|\mathbf{M}|} = \begin{pmatrix} -\frac{1}{6} & \frac{3}{6} & \frac{1}{6} \\ -\frac{10}{6} & -\frac{6}{6} & \frac{4}{6} \\ \frac{9}{6} & \frac{3}{6} & -\frac{3}{6} \end{pmatrix} = \begin{pmatrix} -\frac{1}{6} & \frac{1}{2} & \frac{1}{6} \\ -\frac{5}{3} & -1 & \frac{2}{3} \\ \frac{3}{2} & \frac{1}{2} & -\frac{1}{2} \end{pmatrix}$$

Premultiplying (1) by $\mathbf{M}^{-1}$

$$\begin{pmatrix} -\frac{1}{6} & \frac{1}{2} & \frac{1}{6} \\ -\frac{5}{3} & -1 & \frac{2}{3} \\ \frac{3}{2} & \frac{1}{2} & -\frac{1}{2} \end{pmatrix}\begin{pmatrix} 1 & 2 & 3 \\ 1 & -1 & -1 \\ 4 & 5 & 6 \end{pmatrix}\begin{pmatrix} x \\ y \\ z \end{pmatrix} = \begin{pmatrix} -\frac{1}{6} & \frac{1}{2} & \frac{1}{6} \\ -\frac{5}{3} & -1 & \frac{2}{3} \\ \frac{3}{2} & \frac{1}{2} & -\frac{1}{2} \end{pmatrix}\begin{pmatrix} 5 \\ 0 \\ 11 \end{pmatrix}$$

$$\begin{pmatrix} 1 & 0 & 0 \\ 0 & 1 & 0 \\ 0 & 0 & 1 \end{pmatrix}\begin{pmatrix} x \\ y \\ z \end{pmatrix} = \begin{pmatrix} -\frac{5}{6} + \frac{11}{6} \\ -\frac{25}{3} + \frac{22}{3} \\ \frac{15}{2} - \frac{11}{2} \end{pmatrix} = \begin{pmatrix} 1 \\ -1 \\ 2 \end{pmatrix}$$

$\boxed{x = 1}$ $\boxed{y = -1}$ $\boxed{z = 2}$

---

6. (a) $\mathbf{y} = \mathbf{A}\mathbf{x}$ therefore $\mathbf{x} = \mathbf{A}^{-1}\mathbf{y}$ $\lambda = 3$

$$\mathbf{A} = \begin{pmatrix} 1 & 1 & 1 \\ 1 & 3 & 6 \\ 1 & 2 & 3 \end{pmatrix} \quad \mathbf{y} = \begin{pmatrix} 1 \\ 2 \\ 1 \end{pmatrix}$$

$$|\mathbf{A}| = \begin{vmatrix} 1 & 0 & 0 \\ 1 & 2 & 5 \\ 1 & 1 & 2 \end{vmatrix} = (2 \times 2 - 5 \times 1) = -1 \text{ expanding along first row}$$

$$\text{where } \mathbf{A} = \begin{pmatrix} 1 & 1 & 1 \\ 1 & 3 & 6 \\ 1 & 2 & 3 \end{pmatrix} = \begin{pmatrix} 1 & 1-1 & 1-1 \\ 1 & 3-1 & 6-1 \\ 1 & 2-1 & 3-1 \end{pmatrix} = \begin{pmatrix} 1 & 0 & 0 \\ 1 & 2 & 5 \\ 1 & 1 & 2 \end{pmatrix}$$

using (i) column 2 – column 1

using (ii) column 3 – column 1

$$\mathbf{A}^* = \begin{pmatrix} -3 & 3 & -1 \\ -1 & 2 & -1 \\ 3 & -5 & 2 \end{pmatrix} \qquad \mathbf{A}^{*T} \begin{pmatrix} -3 & -1 & 3 \\ 3 & 2 & -5 \\ -1 & -1 & 2 \end{pmatrix}$$

$$\mathbf{A}^{-1} = \frac{\mathbf{A}^{*T}}{|\mathbf{A}|} = \begin{pmatrix} 3 & 1 & -3 \\ -3 & -2 & 5 \\ 1 & 1 & -2 \end{pmatrix}$$

$$\therefore \mathbf{x} = \begin{pmatrix} 3 & 1 & -3 \\ -3 & -2 & 5 \\ 1 & 1 & -2 \end{pmatrix} \begin{pmatrix} 1 \\ 2 \\ 1 \end{pmatrix} = \begin{pmatrix} 2 \\ -2 \\ 1 \end{pmatrix}.$$

(b) The value of λ for which no unique values of $\mathbf{x}$ can be found given $\mathbf{y}$ is when $|\mathbf{A}| = 0$

$$|\mathbf{A}| = \begin{vmatrix} 1 & 0 & 0 \\ 1 & 2 & 5 \\ 1 & 1 & \lambda - 1 \end{vmatrix} = 2(\lambda - 1) - 5 = 2\lambda - 7$$

$$\therefore |\mathbf{A}| = 0 \Rightarrow 2\lambda - 7 = 0 \quad \therefore \lambda = \frac{7}{2}.$$

Then $\mathbf{A}\mathbf{x} = \mathbf{y}$ have one degree of freedom i.e. they are linearly dependent and are equivalent to the equation.

$x_1 + \ \ x_2 + \ \ x_3 = y_1$

$x_1 + 3x_2 + 6x_3 = y_2$

$x_1 + 2x_2 + \frac{7}{2}x_3 = y_3$

Now $2y_3 = y_1 + y_2$.

Let $y_1 = 2a$ and $y_2 = 2b \quad \therefore y_3 = a + b$.

Thus all vectors **x** transform to the vector to the form $\begin{pmatrix} 2a \\ 2b \\ a+b \end{pmatrix}$.

---

7. (i) $$\mathbf{a} \times \mathbf{b} = \begin{vmatrix} \mathbf{i} & \mathbf{j} & \mathbf{k} \\ 2 & -3 & 5 \\ -3 & 0 & -2 \end{vmatrix} = \mathbf{i}(6) - \mathbf{j}(-4+15) + \mathbf{k}(-9)$$

$$= 6\mathbf{i} - 11\mathbf{j} - 9\mathbf{k}$$

$$(6\mathbf{i} - 11\mathbf{j} - 9\mathbf{k}) \times (\mathbf{i} + \mathbf{j} + \mathbf{k}) = \begin{vmatrix} \mathbf{i} & \mathbf{j} & \mathbf{k} \\ 6 & -11 & -9 \\ 1 & 1 & 1 \end{vmatrix}$$

$$= \mathbf{i}\,(-11+9) - \mathbf{j}\,(6+9) + \mathbf{k}\,(6+11) = -2\mathbf{i} - 15\mathbf{j} + 17\mathbf{k}$$

$$\therefore (\mathbf{a} \times \mathbf{b}) \times \mathbf{c} = -2\mathbf{i} - 15\mathbf{j} + 17\mathbf{k}.$$

(ii) $$\mathbf{a}\,.\,\mathbf{b}.\,\mathbf{c} = \big[(2\mathbf{i} - 3\mathbf{j} + 5\mathbf{k})\,.\,(-3i - 2\mathbf{k})\big].\,(\mathbf{i} + \mathbf{j} + \mathbf{k})$$

$$= (-6 - 10)\,.\,(\mathbf{i} + \mathbf{j} + \mathbf{k})$$

$$= -16\,(\mathbf{i} + \mathbf{j} + \mathbf{k})\,.$$

(iii) $$\mathbf{a}.\,(\mathbf{b} \times \mathbf{c}) = \mathbf{a}.\,(2\mathbf{i} + \mathbf{j} - 3\mathbf{k})$$

$$= (2\mathbf{i} - 3\mathbf{j} + 5\mathbf{k})\,.\,(2\mathbf{i} + \mathbf{j} - 3\mathbf{k})$$

$$= 4 - 3 - 15 = -14$$

where

$$\mathbf{b} \times \mathbf{c} = \begin{vmatrix} \mathbf{i} & \mathbf{j} & \mathbf{k} \\ -3 & 0 & -2 \\ 1 & 1 & 1 \end{vmatrix} = \mathbf{i}\,2 - \mathbf{j}\,(-3+2) + \mathbf{k}\,(-3)$$

$$= 2\mathbf{i} + \mathbf{j} - 3\mathbf{k}.$$

---

8. (a) $\sinh x = 2\sinh\frac{x}{2}\cosh\frac{x}{2} = \frac{2\sinh\frac{x}{2}\cosh\frac{x}{2}}{\cosh^2\frac{x}{2} - \sinh^2\frac{x}{2}}$

dividing Numerator and Denominator by $\cosh^2\frac{x}{2}$

$$\sinh x = \frac{\frac{2\sinh\frac{x}{2}}{\cosh\frac{x}{2}}}{1 - \tanh^2\frac{x}{2}} = \frac{2\tanh\frac{x}{z}}{1 - \tanh^2\frac{x}{2}} = \frac{2t}{1 - t^2}.$$

(b)

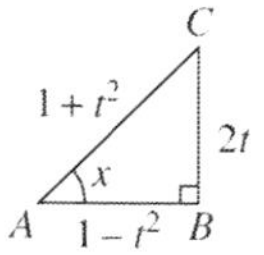

C, $1 + t^2$, $2t$, x, A, $1 - t^2$, B

$\sin x = \frac{2t}{1 + t^2}$ using Pythagoras

$AB^2 = AC^2 - BC^2 \Rightarrow AB = \sqrt{(1 + t^2)^2 - 4t^2} = 1 - t^2$

$$\cos x = \frac{1 - t^2}{1 + t^2} \quad \tan x = \frac{2t}{1 - t^2}, \quad \sec x = \frac{1 + t^2}{1 - t^2}$$

$$\cot x = \frac{1 - t^2}{2t} \quad \text{cosec}\, x = \frac{1 + \text{t}^2}{1 - \text{t}^2}.$$

The corresponding hyperbolic functions are:

$$\cosh x = \frac{1 + t^2}{1 - t^2} \quad \tanh x = \frac{2t}{1 + t^2}, \quad \text{sech}\, x = \frac{1 - t^2}{1 + t^2}$$

$$\coth x = \frac{1 + t^2}{2t} \quad \text{cosech}\, x = \frac{1 - t^2}{1 + t^2}.$$

Examination Test Papers.

GCE Advanced Level.

Further Pure Mathematics FP3

Test Paper 2

Time: 1 hour and 30 minutes.

Instructions and Information

Candidates may use any calculator allowed by the regulations of their Examination Board.

Full marks are awarded for correct answers to ALL questions.

This paper has eight questions.

You can start working with any question and you must label clearly all parts.

1. (a) Sketch (i) $\sinh x$ (ii) $\cosh x$ (iii) $\tanh x$

(iv) $\sinh^{-1} x$ (v) $\cosh^{-1} x$ (vi) $\tanh^{-1} x$. **(6)**

(b) Sketch (i) $\operatorname{cosech} x$ (ii) $\operatorname{sech} x$ (iii) $\coth x$

(iv) $\operatorname{cosech}^{-1} x$ (v) $\operatorname{sech}^{-1} x$ (vi) $\coth^{-1} x$. **(6)**

---

2. Determine the tangent to the ellipse $\frac{x^2}{a^2} + \frac{y^2}{b^2} = 1$ at the point P (a $\cos\theta$, $b \sin\theta$).

Show that the normal at $P(a\cos\theta, b\sin\theta)$ is given by

$yb\cos\theta = ax\sin\theta + \sin\theta\cos\theta(b^2 - a^2)$. **(8)**

---

3. Determine the reduction formula

$I_n\,(n-1) = (n-1)\,I_{n-2} - \tanh x \tanh^{n-2} x$

for $I_n = \int \tanh^n x \, dx$. **(6)**

---

4. Sketch the hyperbolic functions $f(x) = \cosh x$ and $f^{-1}(x) = \cosh^{-1} x$

on the same graph, for $x \geq 0$ former and $x \geq 1$ for the latter.

Find the area under the graph of $f^{-1}(x)$ between $x = 1$ and $x = 2$. **(8)**

---

5. Solve the simultaneous equations

$$\begin{aligned} 2x - 3y + z + 1 &= 0 \\ x + 2y - 2z - 8 &= 0 \\ 3x - 8y + 3z + 8 &= 0 \end{aligned}$$

using matrices. **(15)**

---

6. Show that 9 is an eigenvalue of the matrix

$$\begin{pmatrix} 6 & -2 & 2 \\ -2 & 5 & 0 \\ 2 & 0 & 7 \end{pmatrix}.$$

Find the other two eigenvalues. **(12)**

---

7. Find the angle between the position vector $\overrightarrow{OP} = 3\mathbf{i} + 2\mathbf{j} + \mathbf{k}$

and $\overrightarrow{OQ} = -\mathbf{i} + \mathbf{j} + 4\mathbf{k}$ using the vector product formula. **(9)**

---

8. Find the indefinite integral $\displaystyle\int \frac{\mathrm{d}x}{1 - \sinh x}$. **(5)**

TOTAL FOR PAPER: 75 MARKS

Examination Test Papers.

GCE Advanced Level.

Test Paper 2 Solutions

Further Pure Mathematics FP3

1.

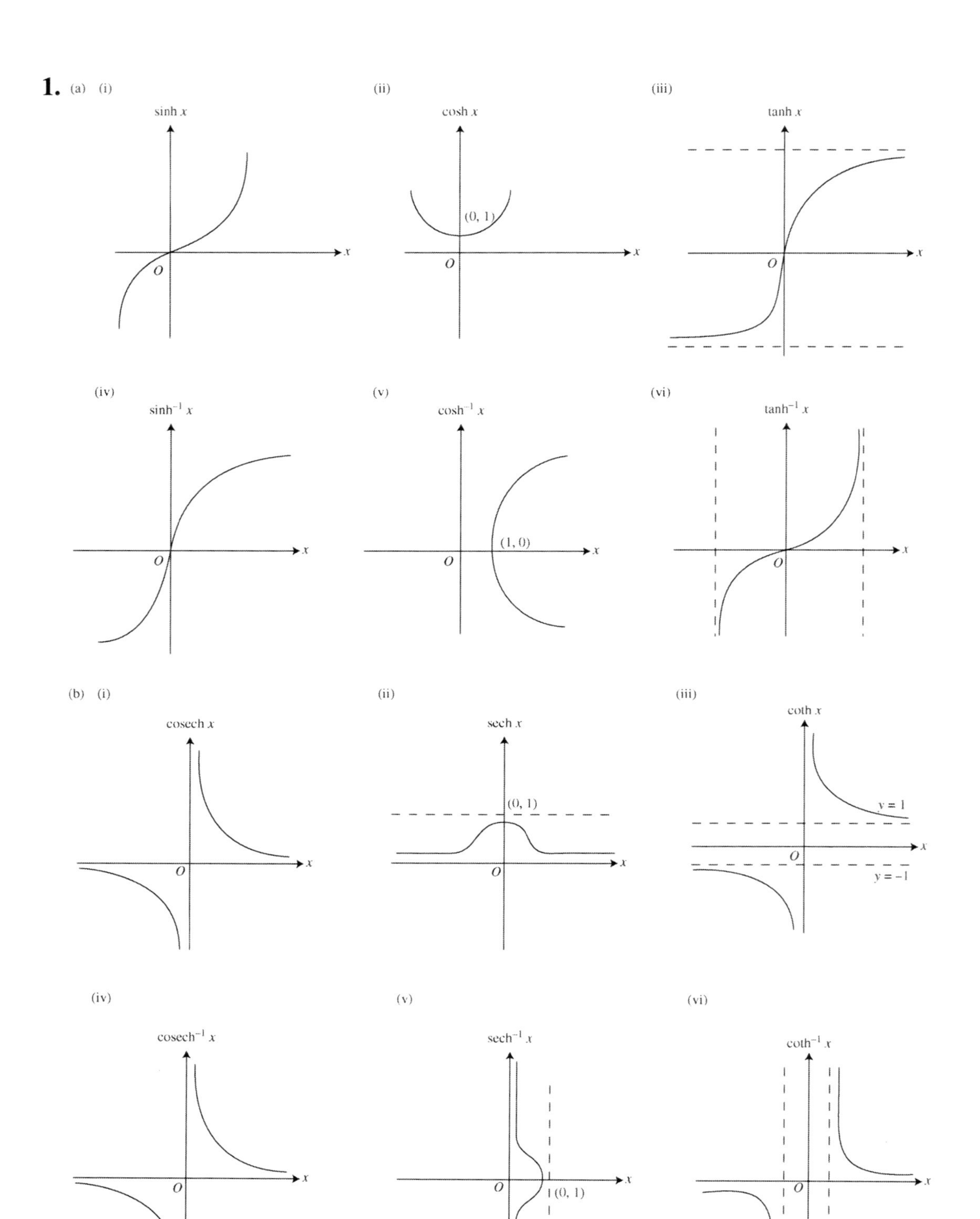
(a) (i)
sinh x
O
x
(ii)
cosh x
(0, 1)
O
x
(iii)
tanh x
O
x
(iv)
sinh⁻¹ x
O
x
(v)
cosh⁻¹ x
(1, 0)
O
x
(vi)
tanh⁻¹ x
O
x
(b) (i)
cosech x
O
x
(ii)
sech x
(0, 1)
O
x
(iii)
coth x
y = 1
O
x
y = −1
(iv)
cosech⁻¹ x
O
x
(v)
sech⁻¹ x
O
(0, 1)
x
(vi)
coth⁻¹ x
O
x
x = −1
x = 1

2. $x = a\cos\theta \qquad \dfrac{dx}{d\theta} = -a\sin\theta \qquad y = b\sin\theta \qquad \dfrac{dy}{d\theta} = b\cos\theta$

$\dfrac{dy}{dx} = \dfrac{b\cos\theta}{-a\sin\theta} = -\dfrac{b}{a}\cot\theta$ the gradient at P.

$y = mx + c \Rightarrow y = \dfrac{b}{a}\cot\theta\, x + c$, this passes through the point P,

$b\sin\theta = -\dfrac{b}{a}\cot\theta\,(a\cos\theta) + c \qquad c = b\sin\theta + b\,\dfrac{\cos^2\theta}{\sin\theta}$

$c = \dfrac{b\sin^2\theta + b\cos^2\theta}{\sin\theta} = \dfrac{b}{\sin\theta} \qquad \therefore y = -\dfrac{b}{a}\cot\theta\, x + \dfrac{b}{\sin\theta}$

$$\boxed{ay\sin\theta + bx\cos\theta = ab}$$

The gradient of the normal at P is $\dfrac{a\tan\theta}{b}$

$y = \dfrac{a\tan\theta}{b}x + c$

$b\sin\theta = \dfrac{a\tan\theta}{b}a\cos\theta + c \qquad b\sin\theta - \dfrac{a^2\sin\theta}{b} = c \Rightarrow c = \dfrac{(b^2 - a^2)}{b}\sin\theta$

$y = \dfrac{a\tan\theta}{b}x + \dfrac{(b^2 - a^2)\sin\theta}{b}$

$$\boxed{by\cos\theta = ax\sin\theta + \sin\theta\cos\theta\,(b^2 - a^2)}$$

---

3. $I_n = \displaystyle\int \tanh^n x\, dx = \int \tanh^{n-2} x \tanh^2 x\, dx = \int \tanh^{n-2} x\left(1 - \operatorname{sech}^2 x\right) dx$

$= \displaystyle\int \underset{②}{\tanh^{n-2} x}\, \underset{①}{dx} - \int \underset{②}{\tanh^{n-2} x}\, \underset{①}{\operatorname{sech}^2 x\, dx}$

$= I_{n-2} - \tanh x \tanh^{n-2} x + \displaystyle\int \tanh x\,(n-2)\tanh^{n-3} x\left(\operatorname{sech}^2 x\right) dx$

$= I_{n-2} - \tanh \tanh^{n-2} x + (n-2)\displaystyle\int \tanh^{n-2} x\left(1 - \tanh^2 x\right) dx$

$= I_{n-2} - \tanh x \tanh^{n-2} x + (n-2)\displaystyle\int \tan h^{n-2} x\, dx - (n-2)\, I_n$

$I_n\,(1 + n - 2) = I_{n-2} - \tanh x \tanh^{n-2} x + (n-2)\, I_{n-2}$

$$\therefore \boxed{I_n\,(n-1) = (n-1)\, I_{n-2} - \tanh x \tanh^{n-2} x}$$

4.

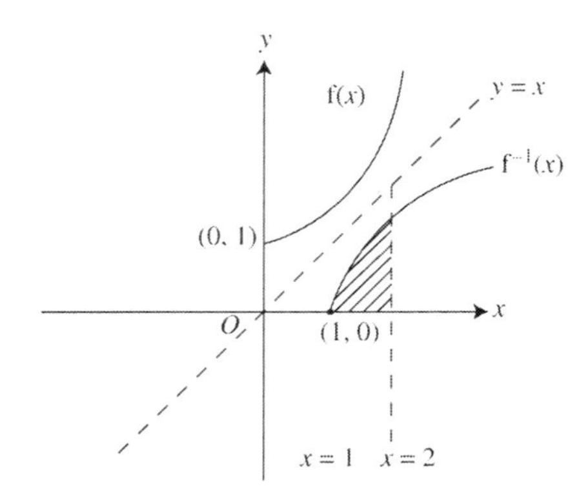

f(x) and $f^{-1}(x)$ are reflections to each other about $y = x$.

$$\int_1^2 f^{-1}(x)\,dx = \int_1^2 \underset{②}{\cosh^{-1}x}\,\underset{①}{dx}$$

using integration by parts

$$= \left[x\cosh^{-1}x\right]_1^2 - \int_1^2 x \times \frac{1}{\sqrt{x^2-1}}\,dx$$

$$= \left[2\cosh^{-1}2 - \cosh^{-1}1\right] - \int_1^2 \frac{1}{2}\,\frac{d\left(x^2-1\right)}{\left(x^2-1\right)^{\frac{1}{2}}}$$

$$= \left[2\,(1.316957897) - 0\right] - \frac{1}{2}\left[\frac{\left(x^2-1\right)^{\frac{1}{2}}}{\frac{1}{2}}\right]_1^2$$

$$= 2.633915794 - \sqrt{3} = 2.633915794 - 1.732050808$$

$$= 0.901864986 = 0.902 \text{ to 3 s.f.}$$

---

5. The three equations can be written in the matrix form

$$\begin{pmatrix} 2 & -3 & 1 \\ 1 & 2 & -2 \\ 3 & -8 & 3 \end{pmatrix}\begin{pmatrix} x \\ y \\ z \end{pmatrix} = \begin{pmatrix} -1 \\ 8 \\ -8 \end{pmatrix}$$

$$\mathbf{M} = \begin{pmatrix} 2 & -3 & 1 \\ 1 & 2 & -2 \\ 3 & -8 & 3 \end{pmatrix} \quad \mathbf{M}^* = \begin{pmatrix} -10 & -9 & -14 \\ 1 & 3 & 7 \\ 4 & 5 & 7 \end{pmatrix}$$

$$\mathbf{M}^{*\mathrm{T}} = \begin{pmatrix} -10 & 1 & 4 \\ -9 & 3 & 5 \\ -14 & 7 & 7 \end{pmatrix} \qquad |\mathbf{M}| = 2(-10) + 3(9) - 14 = -7$$

$$\mathbf{M}^{-1} = \frac{\mathbf{M}^{*\mathrm{T}}}{|\mathbf{M}|} = \begin{pmatrix} \frac{10}{7} & -\frac{1}{7} & -\frac{4}{7} \\ \frac{9}{7} & -\frac{3}{7} & -\frac{5}{7} \\ 2 & -1 & -1 \end{pmatrix}$$

$$\mathbf{I}\begin{pmatrix} x \\ y \\ z \end{pmatrix} = \begin{pmatrix} \frac{10}{7} & -\frac{1}{7} & -\frac{4}{7} \\ \frac{9}{7} & -\frac{3}{7} & -\frac{5}{7} \\ 2 & -1 & -1 \end{pmatrix}\begin{pmatrix} -1 \\ 8 \\ -8 \end{pmatrix} = \begin{pmatrix} -\frac{10}{7} & -\frac{8}{7} & +\frac{32}{7} \\ -\frac{9}{7} & -\frac{24}{7} & +\frac{40}{7} \\ -2 & -8 & +8 \end{pmatrix} = \begin{pmatrix} 2 \\ 1 \\ -2 \end{pmatrix}$$

$$\boxed{x = 2} \quad \boxed{y = 1} \quad \boxed{z = -2}$$

---

6. $\mathbf{A} = \begin{pmatrix} 6 & -2 & 2 \\ -2 & 5 & 0 \\ 2 & 0 & 7 \end{pmatrix}$.

The eigenvalues of **A** are given by the characteristic equation

$\mathbf{A} - \lambda\mathbf{I} = 0$

$$|\mathbf{A} - \lambda\,\mathbf{I}| = \begin{vmatrix} 6-\lambda & -2 & 2 \\ -2 & 5-\lambda & 0 \\ 2 & 0 & 7-\lambda \end{vmatrix}.$$

Put $\lambda = 9 \quad \therefore |\mathbf{A} - 9\mathbf{I}| \begin{vmatrix} -3 & -2 & 2 \\ -2 & -4 & 0 \\ 2 & 0 & -2 \end{vmatrix}$

Now Row 1 + Row 3 $= (-1 \quad -2 \quad 0) = \frac{1}{2}$ Row 2

$\therefore$ rows are linearly dependent $\Rightarrow |\mathbf{A} - 9\mathbf{I}| = 0$

$\therefore$ 9 is an eigenvalue of **A**.

$$|\mathbf{A} - \lambda\mathbf{I}| = 2\begin{vmatrix} 2 & 5-\lambda \\ 2 & 0 \end{vmatrix} + (7-\lambda)\begin{vmatrix} 6-\lambda & -2 \\ -2 & 5-\lambda \end{vmatrix}$$

$$= 2\big[2 \times 0 - 2 \times (5 - \lambda)\big] + (7 - \lambda)\big[(6-\lambda)(5 - \lambda) - 4\big]$$

$$= -20 + 4\lambda + (7 - \lambda)\big[(30 - 11\lambda + \lambda^2) - 4\big]$$

$$= -20 + 4\lambda + (7 - \lambda)(26 - 11\lambda + \lambda^2) = -(\lambda^3 - 18\lambda^2 + 99\lambda - 162).$$

Since $\lambda = 9$ is a root of $(\mathbf{A} - \lambda\mathbf{I}) = 0$ then

$(\lambda - 9)$ is a factor of $|\mathbf{A} - \lambda\mathbf{I}|$

$$\begin{array}{r|l} & -\lambda^2 + 9\lambda - 18 \\ \hline \lambda - 9 & -\lambda^3 + 18\lambda^2 - 99\lambda + 162 \\ & \underline{-\lambda^3 + \ \ 9\lambda^2 \qquad\qquad\qquad} \\ & \qquad 9\lambda^2 - 99\lambda + 162 \\ & \qquad \underline{9\lambda^2 - 81\lambda \qquad\quad} \\ & \qquad\qquad -18\lambda + 162 \\ & \qquad\qquad \underline{-18\lambda + 162} \\ & \qquad\qquad\qquad 0 \end{array}$$

$$\therefore (\lambda - 9)(-\lambda^2 + 9\lambda - 18) = -\lambda^3 + 18\lambda^2 - 99\lambda + 162$$

$$= -(\lambda - 9)(\lambda - 6)(\lambda - 3)$$

$\therefore |\mathbf{A} - \lambda\mathbf{I}| = 0 \Rightarrow \lambda = 3, \lambda = 6, \lambda = 9.$

For $\lambda = 3 \qquad (\mathbf{A} - \lambda\mathbf{I})\, x = 0$

$$3\mathbf{x}_1 - 2\mathbf{x}_2 + 2\mathbf{x}_3 \qquad \ldots \text{(i)}$$

$$-2\mathbf{x}_1 + 2\mathbf{x}_2 = 0 \qquad \ldots \text{(ii)}$$

$$2\mathbf{x}_1 + 4\mathbf{x}_3 = 0 \qquad \ldots \text{(iii)}$$

(i), (ii) and (iii) are linearly dependend. From (ii) $\mathbf{x}_1 : \mathbf{x}_2 = 1 : 1$ (iii) $\mathbf{x}_1 : \mathbf{x}_3 = 2 : -1$

$\therefore \mathbf{x}_1 : \mathbf{x}_2 : \mathbf{x}_3 = 2 : 2 : 1.$

Normalised $\mathbf{x} = \begin{pmatrix} \frac{2}{3} \\ \frac{2}{3} \\ -\frac{1}{3} \end{pmatrix}.$

For $\lambda = 6 \quad (\mathbf{A} + \lambda\mathbf{I})\mathbf{x} = 0$

$$-2\mathbf{x}_2 + 2\mathbf{x}_3 = 0 \qquad \ldots \text{(iv)}$$

$$-2\mathbf{x}_1 - \mathbf{x}_2 = 0 \qquad \ldots \text{(v)}$$

$$2\mathbf{x}_1 + \mathbf{x}_3 = 0 \qquad \ldots \text{(vi)}$$

(v) $\mathbf{x}_1 : \mathbf{x}_2 = 1 : -2$

(vi) $\mathbf{x}_1 : \mathbf{x}_3 = 1 : -2$

$$\mathbf{x}_1 : \mathbf{x}_2 : \mathbf{x}_3 = 1 : -2 : -2.$$

$$\text{Normalised } \mathbf{x} = \begin{pmatrix} \frac{1}{3} \\ -\frac{2}{3} \\ -\frac{2}{3} \end{pmatrix}.$$

For $\lambda = 9 \qquad (\mathbf{A} - \lambda\mathbf{I})\mathbf{x} = 0$

$$-3\mathbf{x}_1 - 2\mathbf{x}_2 + 2\mathbf{x}_3 = 0 \qquad \ldots \text{(vii)}$$

$$-2\mathbf{x}_1 - 4\mathbf{x}_2 = 0 \qquad \ldots \text{(viii)}$$

$$-2\mathbf{x}_1 - 2\mathbf{x}_3 = 0 \qquad \ldots \text{(iv)}$$

(viii) $\Rightarrow \mathbf{x}_1 : \mathbf{x}_2 = 2 : -1$

(ix) $\Rightarrow \mathbf{x}_1 : \mathbf{x}_3 = 1 : 1$

$\therefore \mathbf{x}_1 : \mathbf{x}_2 : \mathbf{x}_3 = 2 : -1 : 2.$

$$\text{Normalised } \mathbf{x} = \begin{pmatrix} \frac{2}{3} \\ -\frac{1}{3} \\ \frac{2}{3} \end{pmatrix}.$$

$$\mathbf{P} = \begin{pmatrix} \frac{2}{3} & \frac{1}{3} & \frac{2}{3} \\ \frac{2}{3} & -\frac{2}{3} & -\frac{1}{3} \\ -\frac{1}{3} & -\frac{1}{3} & \frac{2}{3} \end{pmatrix} \qquad \mathbf{P}^{\mathrm{T}} = \begin{pmatrix} \frac{2}{3} & \frac{2}{3} & -\frac{1}{3} \\ \frac{1}{3} & -\frac{2}{3} & -\frac{1}{3} \\ \frac{2}{3} & -\frac{1}{3} & \frac{2}{3} \end{pmatrix}$$

$$\mathbf{PP}^{\mathrm{T}} = \begin{pmatrix} 1 & 0 & 0 \\ 0 & 1 & 0 \\ 0 & 0 & 1 \end{pmatrix} = \mathbf{I} \qquad \therefore \mathbf{P} \text{ is orthogonal matrix.}$$

7.

y

$P\,(3, -2, 1)$

$Q\,(-1, 1, 4)$

θ

O x

$$\frac{\left|\overrightarrow{OP} \times \overrightarrow{OQ}\right|}{\left|\overrightarrow{OP}\right|\,\left|\overrightarrow{OQ}\right|} = \sin\theta$$

$$\overrightarrow{OP} \times \overrightarrow{OQ} = (3\mathbf{i} - 2\mathbf{j} + \mathbf{k}) \times (-\mathbf{i} + \mathbf{j} + 4\mathbf{k})$$

$$= \begin{vmatrix} \mathbf{i} & \mathbf{j} & \mathbf{k} \\ 3 & -2 & 1 \\ -1 & 1 & 4 \end{vmatrix} = \mathbf{i}\,(-8 - 1) - \mathbf{j}\,(12 + 1) + \mathbf{k}\,(3 - 2)$$

$$= -9\mathbf{i} - 13\mathbf{j} + \mathbf{k}$$

$$\left|\overrightarrow{OP} \times \overrightarrow{OQ}\right| = \sqrt{9^2 + 13^2 + 1^2} = 15.8 \text{ to 3 s.f}$$

$$\left|\overrightarrow{OP}\right| = \sqrt{3^2 + 2^2 + 1^2} = 3.74 \text{ to 3 s.f}$$

$$\left|\overrightarrow{OQ}\right| = \sqrt{1^2 + 1^2 + 4^2} = 4.24 \text{ to 3 s.f}$$

$$\sin\theta = \frac{15.8}{3.74 \times 4.24} = 0.996367672 \qquad \theta = 85.1°.$$

Check that when rounding off only at the end, $\theta = 86.4°$.

---

8. $\displaystyle\int \frac{\mathrm{d}x}{i - \sinh x}$ let $t = \tanh\dfrac{x}{2} \Rightarrow \dfrac{\mathrm{d}t}{\mathrm{d}x} = \operatorname{sech}^2\dfrac{x}{2} \Rightarrow \mathrm{d}x = \dfrac{\mathrm{d}t}{\operatorname{sech}^2\dfrac{x}{2}}$

$$\mathrm{d}x = \frac{\mathrm{d}t}{1 - \tanh^2\dfrac{x}{2}} = \frac{\mathrm{d}t}{1 - t^2}$$

$$\int \frac{\frac{\mathrm{d}t}{1-t^2}}{1 - \frac{2t}{1-t^2}} = \int \frac{\mathrm{d}t}{1 - t^2 - 2t} = -\int \frac{\mathrm{d}t}{t^2 + 2t - 1} = -\int \frac{\mathrm{d}x}{(t+1)^2 - 2}$$

$$\frac{1}{(t + 1^2) - \left(\sqrt{2}\right)^2} = \frac{1}{\left(t + 1 - \sqrt{2}\right)\left(t + 1 + \sqrt{2}\right)} \equiv \frac{A}{t + 1 - \sqrt{2}} + \frac{B}{t + 1 + \sqrt{2}}$$

$$1 \equiv A\left(t + 1 + \sqrt{2}\right) + B\left(t + 1 - \sqrt{2}\right)$$

if $t + 1 = -\sqrt{2}$ $\quad \mathrm{B} = \dfrac{1}{-2\sqrt{2}} = -\dfrac{1}{2\sqrt{2}}$

if $t + 1 = \sqrt{2}$ $\quad A = \dfrac{1}{2\sqrt{2}}$

$$-\int \frac{\mathrm{d}t}{(t+1)^2-2} = -\int \frac{1\,\mathrm{d}t}{2\sqrt{2}\left[t+1-\sqrt{2}\right]} + \int \frac{\mathrm{d}t}{2\sqrt{2}\left[t+1+\sqrt{2}\right]}$$

$$= \frac{1}{2\sqrt{2}} \ln \left[t+1+\sqrt{2}\right] - \frac{1}{2\sqrt{2}} \ln \left[t+1-\sqrt{2}\right] + \ln A$$

$$= \frac{1}{2\sqrt{2}} \ln A \frac{(t+1+\sqrt{2})}{(t+1-\sqrt{2})}.$$

Examination Test Papers.

GCE Advanced Level.

Further Pure Mathematics FP3

Test Paper 3

Time: 1 hour and 30 minutes.

Instructions and Information

Candidates may use any calculator allowed by the regulations of their Examination Board.

Full marks are awarded for correct answers to ALL questions.

This paper has eight questions.

You can start working with any question and you must label clearly all parts.

1. Prove

(i) $\sinh^{-1} x = \ln\left(x + \sqrt{x^2}\right)$

(ii) $\cosh^{-1} x = \pm \ln\left(x + \sqrt{x^2 - 1}\right) \quad (x \geq 1)$

(iii) $\tanh^{-1} x = \frac{1}{2} \ln \frac{1 + x}{1 - x} \quad (|x| < 1)$. **(12)**

---

2. Sketch the ellipse $\frac{x^2}{a^2} + \frac{y^2}{b^2} = 1 \quad a > b > 0$ showing the foci, the directrices and the coordinates of the curve intersecting the axes.

Write down the relationship between a, b and e.

If the eccentricity of an ellipse is $e > \frac{2}{3}$ and $a = 5$, determine the value of b and hence write down the equation of the curve.

Find the coordinates of the foci and the equations of the directrices. **(10)**

---

3. If $I_n = \int_0^{\frac{\pi}{4}} \sin^n x \, dx$ show that

$$I_n = -\tfrac{1}{n}\left(\tfrac{1}{\sqrt{2}}\right)^n + \tfrac{n-1}{n}\left(-\tfrac{1}{n-2}\right)\left(\tfrac{1}{\sqrt{2}}\right)^{n-2} + \left(\tfrac{n-3}{n-2}\right)\left(-\tfrac{1}{n+4}\right)\left(\tfrac{1}{\sqrt{2}}\right)^{n-4}$$

$+\frac{n-5}{n-4} I_{n-6}$ and so on. **(10)**

---

4. Sketch the exponential functions $f(x) = e^x$, $-g(x)$, $g(x) = e^{-x}$ on the same graph.

Hence sketch the curves of $\frac{e^x + e^{-x}}{2}$ and $\frac{e^x - e^{-x}}{2}$ on two separate graphs, and write down these two new functions. **(6)**

---

5. If $\mathbf{A} = \begin{pmatrix} 3 & -1 & 2 \\ -1 & 2 & 0 \\ 4 & -5 & 7 \end{pmatrix}$ and $\mathbf{B} = \begin{pmatrix} 0 & 1 & 2 \\ 3 & 0 & 4 \\ 5 & 6 & 0 \end{pmatrix}$.

Find **AB** and **BA**. **(6)**

---

6. Find the eigenvalues of the matrix **A** where

$$\mathbf{A} = \begin{pmatrix} 2 & 2 & 1 \\ 1 & 3 & 1 \\ 1 & 2 & 2 \end{pmatrix}.$$

Find an eigenvector corresponding to the non repeated eigen value. **(12)**

---

7. Find the perpendicular distance of the point $(-1, 2, -4)$ from the line whose vector equations is given $\mathbf{r} = (2\mathbf{i} + \mathbf{j} + \mathbf{k}) + \lambda\,(-3\mathbf{i} + 4\mathbf{j} - 5\mathbf{k})$. **(10)**

---

8. Show that the derivatives of $\sinh^{-1} x$ and $\text{cosech}^{-1} x$ are $\dfrac{1}{\sqrt{1+x^2}}$ and $-\dfrac{1}{x\sqrt{1+x^2}}$ respectively. **(9)**

TOTAL FOR PAPER: 75 MARKS

Examination Test Papers.

GCE Advanced Level.

Test Paper 3 Solutions

Further Pure Mathematics FP3

1. (i) $y = \sinh^{-1} x$

$$x = \sinh y = \frac{1}{2}\left(e^{y} - e^{-y}\right) \qquad e^{y} - e^{-y} = 2x$$

multiplying both sides by e^{y}

$$e^{2y} - 2xe^{y} - 1 = 0 \qquad e^{y} = \frac{2x \pm \sqrt{4x^2+4}}{2} = x \pm \sqrt{x^2+1}$$

taking logarithms to the base e

$$y = \ln\left(x \pm \sqrt{x^2+1}\right) \qquad \ln\left(x - \sqrt{x^2+1}\right) \text{ is } \underline{\text{not}} \text{ defined}$$

$$\boxed{\sinh^{-1} x = \ln\left(x + \sqrt{x^2+1}\right)}$$

(ii) $y = \cosh^{-1} x \Rightarrow x = \cosh y = \frac{1}{2}\left(e^{y} + e^{-y}\right)$

$e^{y} + e^{-y} = 2x$ and multiplying both sides by e^{y}

$$e^{2y} - 2xe^{y} + 1 = 0 \Rightarrow e^{y} = x \pm \sqrt{x^2-1}$$

$$\therefore y = \ln\left(x \pm \sqrt{x^2-1}\right) = \ln\left(x + \sqrt{x^2-1}\right) \text{ or } \ln\left(x - \sqrt{x^2-1}\right)$$

$$\text{but } \frac{1}{x+\sqrt{x^2-1}} = x - \sqrt{x^2-1} \Rightarrow 1 = x^2 - \left(x^2-1\right)$$

$$\therefore y = \ln\left(x + \sqrt{x^2-1}\right) \text{ or } y = \ln\frac{1}{x+\sqrt{x^2-1}}$$

$$\therefore \cosh^{-1} x = \pm\ln\left(x + \sqrt{x^2-1}\right) \quad (x \geq 1)$$

the principal value of $\cosh^{-1} x$ is given by $x = \ln\left(x + \sqrt{x^2-1}\right)$.

(iii) $y = \tanh^{-1} x \Rightarrow x = \tanh y = \dfrac{e^y - e^{-y}}{e^y + e^{-y}} \times \dfrac{e^y}{e^y}$

$$x = \frac{e^{2y} - 1}{e^{2y} + 1} \Rightarrow xe^{2y} + x = e^{2y} - 1$$

$1 + x = e^{2y}(1 - x) \Rightarrow e^{2y} = \dfrac{1 + x}{1 - x}$ taking logarithms to the base on both sides

$$2y = \ln \frac{1 + x}{1 - x} \qquad \therefore \tanh^{-1} x = \frac{1}{2} \ln \frac{1 + x}{1 - x} \; (|x| < 1).$$

2.

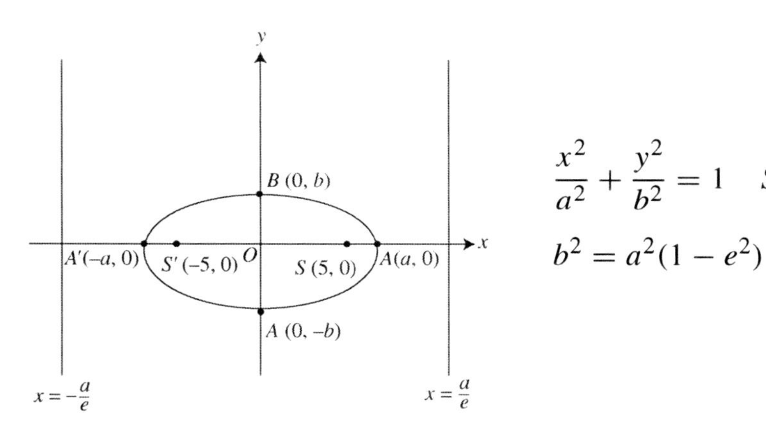

$$\frac{x^2}{a^2} + \frac{y^2}{b^2} = 1 \quad S(ae, 0) \quad S'(-ae, 0)$$

$$b^2 = a^2(1 - e^2)$$

$$b^2 = 5^2\left[1 - \left(\frac{2}{3}\right)^2\right] = 25\left[1 - \frac{4}{9}\right] = 25 \times \frac{5}{9} = \frac{125}{9} \quad b = \pm\frac{5}{3}\sqrt{5} \quad \frac{x^2}{5^2} + \frac{y^2}{\frac{125}{9}} = 1$$

$$\boxed{5x^2 + 9y^2 = 125}$$

$$S(ae, 0) \equiv S\left(5 \times \tfrac{2}{3}, 0\right) = \boxed{S\left(\frac{10}{3}, 0\right)} \quad S'(-ae, 0) \equiv S\left(-5 \times \tfrac{2}{3}, 0\right) = \boxed{S\left(-\frac{10}{3}, 0\right)}$$

$x = \frac{5}{e} = \frac{5}{\frac{2}{3}} = \frac{15}{2}$ and $x = -\frac{5}{e} = -\frac{5}{\frac{2}{3}} = -\frac{15}{3}$

$$\boxed{x = \frac{15}{3}} \qquad \boxed{x = -\frac{15}{3}}$$

---

3. $I_n = \int_0^{\frac{\pi}{4}} \sin^n x \mathrm{d}x = \left[-\frac{1}{n}\cos x \sin^{n-1} x\right]_0^{\frac{\pi}{4}} + \frac{n-1}{n} I_{n-2}$

$$= -\frac{1}{n}\cos\frac{\pi}{4}\sin^{n-1}\frac{\pi}{4} + \frac{n-1}{n} I_{n-2}$$

$$= -\frac{1}{n}\frac{1}{\sqrt{2}}\left(\frac{1}{\sqrt{2}}\right)^{n-1} + \frac{n-1}{n} I_{n-2}$$

$$I_{n-2} = \int_0^{\frac{\pi}{4}} \sin^{n-2} x\,\mathrm{d}x$$

$$= \left[-\frac{1}{n-2}\cos x \sin^{n-3} x\right]_0^{\frac{\pi}{4}} + \frac{n-3}{n-2} I_{n-4} = -\frac{1}{n-2}\frac{1}{\sqrt{2}}\left(\frac{1}{\sqrt{2}}\right)^{n-3} + \frac{n-3}{n-2} I_{n-4}$$

$$I_{n-4} = \int_0^{\frac{\pi}{4}} \sin^{n-4} x\,\mathrm{d}x = \left[-\frac{1}{n-4}\cos x \sin^{n-5} x\right]_0^{\frac{\pi}{4}} + \frac{n-5}{n-4} I_{n-6}$$

$$= -\frac{1}{n-4}\frac{1}{\sqrt{2}}\left(\frac{1}{\sqrt{2}}\right)^{n-5} + \frac{n-5}{n-4} I_{n-6}$$

$$I_n = -\tfrac{1}{n}\left(\tfrac{1}{\sqrt{2}}\right)^n + \tfrac{n-1}{n}\left(-\tfrac{1}{n-2}\right)\left(\tfrac{1}{\sqrt{2}}\right)^{n-2} + \left(\tfrac{n-3}{n-2}\right)\left(-\tfrac{1}{n-4}\right)\left(\tfrac{1}{\sqrt{2}}\right)^{n-4} + \tfrac{n-5}{n-4} I_{n-6}$$

and so on.

4.

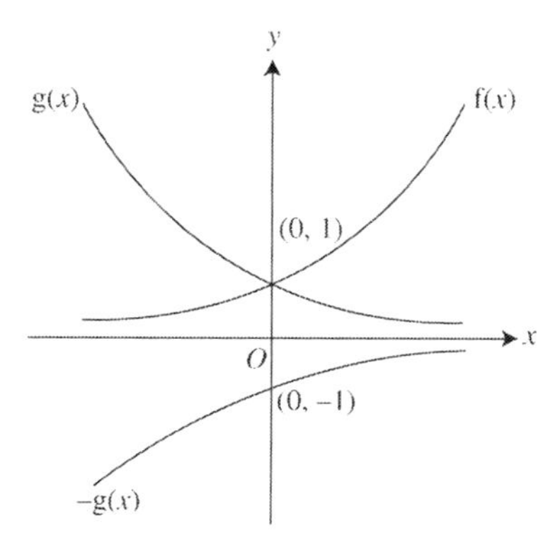

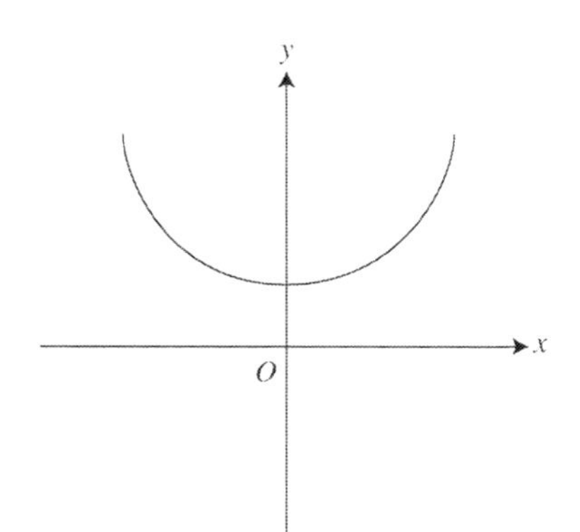

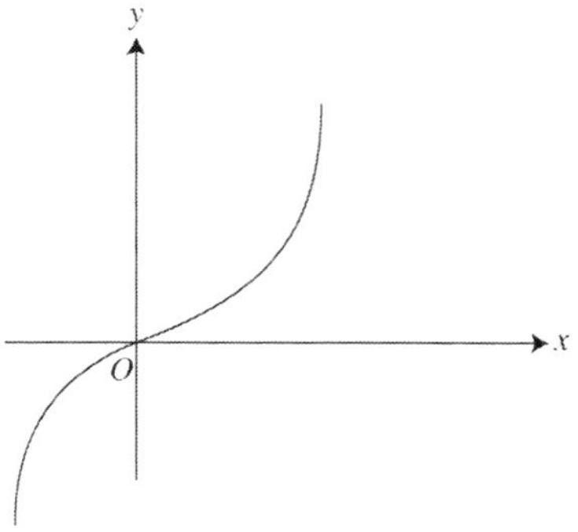

$$\cosh x = \frac{e^x + e^{-x}}{2} \qquad \sinh x = \frac{e^x - e^{-x}}{2}$$

$$= \frac{e^x + \left(-e^{-x}\right)}{2}.$$

---

5. $\mathbf{AB} = \begin{pmatrix} 3 & -1 & 2 \\ -1 & 2 & 0 \\ 4 & -5 & 7 \end{pmatrix} \begin{pmatrix} 0 & 1 & 2 \\ 3 & 0 & 4 \\ 5 & 6 & 0 \end{pmatrix}$

$$= \begin{pmatrix} 3 \times 0 + (-1) \times 3 + 2 \times 5 & 3 \times 1 + (-1) \times 0 + 2 \times 6 & 3 \times 2 + (-1) \times 4 + 2 \times 0 \\ (-1) \times 0 + 2 \times 3 + 0 \times 5 & (-1)1 + 2 \times 0 + 0 \times 5 & (-1) \times 2 + 2 \times 4 + 0 \times 0 \\ 4 \times 0 + -5 \times 3 + 7 \times 5 & 4 \times 1 + (-5) \times 0 + 7 \times 6 & 4 \times 2 + (-5 \times 4) + 7 \times 0 \end{pmatrix}$$

$$= \begin{pmatrix} 7 & 15 & 2 \\ 6 & -1 & 6 \\ 20 & 46 & -12 \end{pmatrix}$$

$\mathbf{BA} = \begin{pmatrix} 0 & 1 & 2 \\ 3 & 0 & 4 \\ 5 & 6 & 0 \end{pmatrix} \begin{pmatrix} 3 & -1 & 2 \\ -1 & 2 & 0 \\ 4 & -5 & 7 \end{pmatrix} = \begin{pmatrix} -1 + 8 & 2 - 10 & 14 \\ 9 + 16 & -3 - 20 & 6 + 28 \\ 15 - 6 & -5 + 12 & 10 \end{pmatrix}$

$$= \begin{pmatrix} 7 & -8 & 14 \\ 25 & -23 & 34 \\ 9 & 7 & 10 \end{pmatrix}$$

$\mathbf{AB} = \mathbf{BA}$ **A** and **B** are not commutative.

---

6. Eigenvalues λ given by $|\mathbf{A} - \lambda\mathbf{I}| = 0$ the charactertic equation

$$|\mathbf{A} - \lambda I| = \begin{vmatrix} 2-\lambda & 2 & 1 \\ 1 & 3-\lambda & 1 \\ 1 & 2 & 2-\lambda \end{vmatrix}$$

$$= \begin{vmatrix} 1-\lambda & -1+\lambda & 0 \\ 0 & 1-\lambda & -1+\lambda \\ 1 & 2 & 2-\lambda \end{vmatrix}$$

using (i) Row 1 − Row 2

and using (ii) Row 2 − Row 3

$$= (1-\lambda)^2 \begin{vmatrix} 1 & -1 & 0 \\ 0 & 1 & -1 \\ 1 & 2 & 2-\lambda \end{vmatrix}$$

$$= (1-\lambda)^2 \begin{vmatrix} 1 & -1 & 0 \\ 0 & 1 & -1 \\ 0 & 3 & 2-\lambda \end{vmatrix}$$

using Row 3 − Row 1

$$= (1-\lambda)^2\big[(1)(2-\lambda) + 3\big] = (1-\lambda)^2(5-\lambda)$$

$$\therefore |\mathbf{A} - \lambda\mathbf{I}| = 0 \Rightarrow \lambda = 1 \text{ twice and } \lambda = 5.$$

For $\lambda = 5$ $\quad |\mathbf{A} - \lambda\mathbf{I}| = 0 \Rightarrow -3\mathbf{x}_1 + 2\mathbf{x}_2 + \mathbf{x}_3 = 0 \qquad \ldots \text{(i)}$

$$\mathbf{x}_1 - 2\mathbf{x}_2 + \mathbf{x}_3 = 0 \qquad \ldots \text{(ii)}$$

$$\mathbf{x}_1 - 2\mathbf{x}_2 - 3\mathbf{x}_3 = 0 \qquad \ldots \text{(iii)}$$

equations (i), (ii) and (iii) are linearly dependent

$\therefore$ (iii) say is redundant

Consider (i) − (ii) $\Rightarrow -4\mathbf{x}_1 + 4\mathbf{x}_2 = 0$

$$\therefore \mathbf{x}_1 : \mathbf{x}_2 = 1{:}1$$

$$\text{(i)} + \text{(ii)} \quad -2\mathbf{x}_1 + 2\mathbf{x}_3 = 0$$

$$\therefore \mathbf{x}_1 : \mathbf{x}_3 = 1{:}1$$

$\therefore$ eigen vector $\mathbf{x} = \begin{pmatrix} 1 \\ 1 \\ 1 \end{pmatrix}$

---

7. $\mathbf{b} = -3\mathbf{i} + 4\mathbf{j} - 5\mathbf{k}$ the direction vector, $\mathbf{p} = -\mathbf{i} + 2\mathbf{j} - 4\mathbf{k}$ the position vector of the point, $\mathbf{a} = 2\mathbf{i} + \mathbf{j} - \mathbf{k}$ the position vector of point A through the line

$$\mathbf{p} - \mathbf{a} = -3\mathbf{i} + \mathbf{j} - 3\mathbf{k}$$

$$|\mathbf{b}| = \sqrt{(-3)^2 + 4^2 + (-5)^2} = \sqrt{50} = 7.07 \text{ to 3 s.f.}$$

$$\mathbf{b} \times (\mathbf{p} - \mathbf{a}) = (-3\mathbf{i} + 4\mathbf{j} - 5\mathbf{k}) \times (-3\mathbf{i} + \mathbf{j} - 3\mathbf{k})$$

$$= \begin{vmatrix} \mathbf{i} & \mathbf{j} & \mathbf{k} \\ -3 & 4 & -5 \\ -3 & 1 & -3 \end{vmatrix} = \mathbf{i}\,(-12 + 5) - \mathbf{j}\,(9 - 15) + \mathbf{k}\,(-3 + 12)$$

$$= -7\mathbf{i} + 6\mathbf{j} + 9\mathbf{k}$$

$$|\mathbf{b} \times (\mathbf{p} - \mathbf{a})| = \sqrt{49 + 36 + 81} = \sqrt{106} = 12.9 \text{ to 3 s.f.}$$

$$\mathbf{d} = \frac{|\mathbf{b} \times (\mathbf{p} - \mathbf{a})|}{|\mathbf{b}|} = \frac{12.9}{7.09} = 1.82 \text{ to 3 s.f.}$$

---

8. $y = \sinh^{-1} x$

$$x = \sinh y \Rightarrow \frac{dx}{dy} = \cosh y \Rightarrow \frac{dy}{dx} = \frac{dy}{dx} = \frac{1}{\cosh y} = \frac{1}{\sqrt{1 + x^2}}$$

$$\cosh^2 y - \sinh^2 y = 1 \quad \cosh y = \sqrt{1 + \sinh^2 y} = \sqrt{1 + x^2} \quad 1 - \coth^2 x = \operatorname{cosech}^2 x$$

$$y = \operatorname{cosech}^{-1} x \qquad x = \operatorname{cosech} y = \frac{1}{\sinh y} \Rightarrow \frac{dx}{dy} = -\frac{1 \times \cosh y}{\sinh^2 y}$$

$$\frac{dy}{dx} = \frac{-\sinh^2 y}{\cosh y} = \frac{-\sinh^2 y}{\sqrt{1 + \sinh^2 y}} = \frac{-\frac{1}{\operatorname{cosech}^2 y}}{\sqrt{1 + \frac{1}{\operatorname{cosech}^2 y}}}$$

$$= -\frac{1}{\operatorname{cosech} y\sqrt{1 + \operatorname{cosech}^2 y}} = \frac{1}{-x\sqrt{1 + x^2}}.$$

Examination Test Papers.

GCE Advanced Level.

Further Pure Mathematics FP3

Test Paper 4

Time: 1 hour and 30 minutes.

Instructions and Information

Candidates may use any calculator allowed by the regulations of their Examination Board.

Full marks are awarded for correct answers to ALL questions.

This paper has eight questions.

You can start working with any question and you must label clearly all parts.

1. Prove

(i) $\operatorname{cosech}^{-1} x = \ln\left(\frac{1}{x} \pm \sqrt{\frac{1+x^2}{x^2}}\right) \quad (|x| > 0)$.

(ii) $\operatorname{sech}^{-1} x = \ln\left(\frac{1}{x} \pm \sqrt{\frac{1-x^2}{x^2}}\right) \quad (0 \le x \le 1)$.

(iii) $\coth^{-1} x = \frac{1}{2}\ln\frac{x+1}{x-1} \qquad |\, x \,| > 1.$ **(10)**

---

2. The foci of an ellipse are $(\pm 4, 0)$ and the eccentricities $e = \pm\frac{4}{5}$.

Find the equation of the ellipse. **(6)**

---

3. If $I_n = \int_0^{\frac{\pi}{2}} \sin^n x \, dx$ show that $I_n = \frac{n-1}{n}\,\frac{n-3}{n-2}\,\frac{n-5}{n-4}\, I_{n-6}$.

If n is even $I_n = \frac{n-1}{n}\,\frac{n-3}{n-2}\,\frac{n-5}{n-4}\,\frac{n-7}{n-6} \ldots \frac{3}{4}, \frac{1}{2} I_0$. **(10)**

---

4. Use the identities

$\cosh(A+B) = \cosh A \cosh B + \sinh A \sinh B$

to express $5 \sinh x + 12 \cosh x$ in the form $R \cosh(x - \alpha)$. **(8)**

---

5. Find the eigenvalues and corresponding eigenvectors of the matrix

$$\mathbf{A} = \begin{pmatrix} 2 & 1 & 1 \\ 1 & 2 & 1 \\ 1 & 1 & 2 \end{pmatrix}.$$ **(12)**

---

6. The matrix $\mathbf{A} = \begin{pmatrix} 2 & 1 & 1 \\ 1 & 2 & 1 \\ 1 & 1 & 2 \end{pmatrix}$.

Find the eigenvalues and hence write down the diagonal matrix. **(8)**

---

7. A line passes through two points $A(1, 2, -3)$ and $B(-2, -3, 1)$.

Find the perpendicular distance from the origin to the line. **(9)**

---

8. (a) Determine the following derivatives:

(i) $y = \text{sech}^2 x$ (ii) $y = \sinh^4 x$ (iii) $y = \sqrt{2}\,\text{cosech}^{\frac{1}{2}} x$.

(b) Differentiate the following functions:-

(i) $y = 3\sinh^{-1}\frac{1}{x}$ (ii) $y = 5\cosh^{-1}(x^2 - 3x + 2)$. **(12)**

TOTAL FOR PAPER: 75 MARKS

Examination Test Papers.

GCE Advanced Level.

Test Paper 4 Solutions

Further Pure Mathematics FP3

1. (i) $y = \text{cosech}^{-1}x \Rightarrow x = \text{cosech}\,y = \dfrac{2}{e^y - e^{-y}}$

$xe^y - xe^{-y} = 2$ and multiplying by e^y $\quad xe^{2y} - 2e^y - x = 0$

$$e^y = \frac{2 \pm \sqrt{4 + 4x^2}}{2x} = \frac{1 \pm \sqrt{1 + x^2}}{x}$$

$$\boxed{\text{cosech}^{-1}x = \ln \frac{1}{x} \pm \sqrt{\frac{1 + x^2}{x}}} \quad |x| > 0.$$

(ii) $y = \text{sech}^{-1}x \Rightarrow x = \text{sech}\,y = \dfrac{2}{e^y + e^{-y}}$

$xe^y + xe^{-y} = 2$ and multiplying by e^y both sides $\quad xe^{2y} - 2e^y + x = 0$

$$e^y = \frac{2 \pm \sqrt{(4 - 4x^2)}}{2x} = \frac{1 \pm \sqrt{1 - x^2}}{x},\ y = \ln\left(\frac{1}{x} \pm \sqrt{\frac{1 - x^2}{x^2}}\right)$$

$$\therefore \boxed{\text{sech}^{-1}x = \ln\left(\frac{1}{x} \pm \sqrt{\frac{1 - x^2}{x^2}}\right)} \quad 0 < x \le 1.$$

(iii) $y = \coth^{-1} x \Rightarrow x = \coth y = \dfrac{e^y + e^{-y}}{e^y - e^{-y}}$

$$e^y x - e^{-y}x = e^y + e^{-y} \Rightarrow e^y x - e^y = xe^{-y} + e^{-y}$$

$$e^{2y}x - e^{2y} = x + 1 \Rightarrow e^{2y} = \frac{x + 1}{x - 1}$$

$$2y = \ln \frac{x + 1}{x - 1} \Rightarrow y = \frac{1}{2} \ln \left|\frac{x + 1}{x - 1}\right|$$

$$\boxed{\coth^{-1} x = \frac{1}{2} \ln \frac{x + 1}{x - 1}} \quad |x| > 1.$$

---

2. $ae = 4e = \dfrac{4}{5} \qquad a = \dfrac{4}{e} = \dfrac{4}{\frac{4}{5}} = 5$

$$b^2 = a^2\left(1 - e^2\right) = 25\left(1 - \frac{16}{25}\right) 25 \times \frac{9}{25} = 9$$

$b = \pm 3$

$$\boxed{\frac{x^2}{5^2} + \frac{y^2}{3^2} = 1}$$

---

3. $I_n = \int_0^{\frac{\pi}{2}} \sin^n x \, dx = \left[-\frac{1}{n} \cos x \sin^{n-1} x \right]_0^{\frac{\pi}{2}} + \frac{n-1}{n} I_{n-2} = \frac{n-1}{2} I_{n-2}$

$$I_{n-2} = \int_0^{\frac{\pi}{2}} \sin^{n-2} x \, dx = \left[-\frac{1}{n-2} \cos x \sin^{n-3} x \right]_0^{\frac{\pi}{2}} + \frac{n-3}{n-2} I_{n-4} = \frac{n-3}{n-2} I_{n-4}$$

$$I_{n-4} = \int_0^{\frac{\pi}{2}} \sin^{n-4} x \, dx = \left[-\frac{1}{n-4} \cos x \sin^{n-5} x \right]_0^{\frac{\pi}{2}} + \frac{n-5}{n-4} I_{n-6}$$

$$I_{n-4} = \frac{n-5}{n-4} I_{n-6}$$

$$I_n = \frac{n-1}{n} \frac{n-3}{n-2} \frac{n-5}{n-4} I_{n-6}.$$

If n is even

$$\boxed{I_n = \frac{n-1}{n} \frac{n-3}{n-2} \frac{n-5}{n-4} \frac{n-7}{n-6} \cdots \frac{3}{4} \frac{1}{2} \frac{\pi}{2}} \text{ where } I_0 = \int_0^{\frac{\pi}{2}} dx = \frac{\pi}{2}.$$

If n is odd

$$\boxed{I_n = \frac{n-1}{n} \frac{n-3}{n-2} \frac{n-5}{n-4} \cdots \frac{4}{5} \cdot \frac{2}{3} \cdot I_1} \text{ where } I_1 = \int_0^{\frac{\pi}{2}} \sin x \, dx = 1.$$

---

4. $R \cosh (x - \alpha) = R \cosh x \cosh \alpha + R \sinh x \sinh \alpha$

$$= \left(\frac{12}{13} \cosh x + \frac{5}{13} \sinh x \right) 13 = 13 \left(\cosh \alpha \cosh x + \sinh \alpha \sinh x \right)$$

$$= 13 \cosh (x - \alpha)$$

$$R = \sqrt{5^2 + 12^2} = 13$$

$$\sin \alpha = \frac{5}{13} \qquad \cos \alpha = \frac{12}{13}$$

$$\alpha = \tanh^{-1} \frac{5}{12} = 0.443651597$$

$\alpha = 0.444$ to 3 s.f.

---

5. The characteristic equation for eigenvalues is $|\mathbf{A} - \lambda \mathbf{I}| = 0$

$$\begin{vmatrix} 2-\lambda & 1 & 1 \\ 1 & 2-\lambda & 1 \\ 1 & 1 & 2-\lambda \end{vmatrix} = 0$$

$$(2-\lambda)\begin{vmatrix} 2-\lambda & 1 \\ 1 & 2-\lambda \end{vmatrix} - \begin{vmatrix} 1 & 1 \\ 1 & 2-\lambda \end{vmatrix} + \begin{vmatrix} 1 & 2-\lambda \\ 1 & 1 \end{vmatrix} = 0$$

$$(2-\lambda)\left[(2-\lambda)^2 - 1\right] - (2-\lambda-1) + (1-2+\lambda) = 0$$

$$(2-\lambda)\left(4 - 4\lambda + \lambda^2 - 1\right) - (1-\lambda) + (-1+\lambda) = 0$$

$$(2-\lambda)(1-\lambda)(3-\lambda) - 2 + 2\lambda = 0$$

$$(2-\lambda)(1-\lambda)(3-\lambda) + 2(\lambda - 1) = 0$$

$$(1-\lambda)(-3\lambda - 2\lambda + \lambda^2) = (1-\lambda)(\lambda^2 - 5\lambda + 4)$$

$$(1-\lambda)(\lambda-4)(\lambda-1) = 0$$

$\lambda = 1, \lambda = 4, \lambda = 1.$

To find the eigenvectors, we solve $(\mathbf{A} - \lambda\mathbf{I})\,\mathbf{x} = \mathbf{0}$

$$\begin{pmatrix} 2-\lambda & 1 & 1 \\ 1 & 2-\lambda & 1 \\ 1 & 1 & 2-\lambda \end{pmatrix}\begin{pmatrix} x_1 \\ x_2 \\ x_3 \end{pmatrix} = \begin{pmatrix} 0 \\ 0 \\ 0 \end{pmatrix}$$

For $\lambda = 1$ $\begin{pmatrix} 1 & 1 & 1 \\ 1 & 1 & 1 \\ 1 & 1 & 1 \end{pmatrix}\begin{pmatrix} x_1 \\ x_2 \\ x_3 \end{pmatrix} = \begin{pmatrix} 0 \\ 0 \\ 0 \end{pmatrix}$

$x_1 + x_2 + x_3 = 0$

$x_1 + x_2 + x_3 = 0$

$x_1 + x_2 + x_3 = 0.$

For $\lambda = 4$ $\begin{pmatrix} -2 & 1 & 1 \\ 1 & -2 & 1 \\ 1 & 1 & -2 \end{pmatrix}\begin{pmatrix} x_1 \\ x_2 \\ x_3 \end{pmatrix} = \begin{pmatrix} 0 \\ 0 \\ 0 \end{pmatrix}$

$$-2x_1 + x_2 + x_3 = 0 \quad \ldots(1)$$
$$x_1 - 2x_2 + x_3 = 0 \quad \ldots(2)$$
$$x_1 + x_2 - 2x_3 = 0. \quad \ldots(3)$$

$$(2) - (3) \qquad -3x_2 + 3x_3 = 0 \Rightarrow x_2 = x_3 = k$$
$$x_1 - 2k + k = 0 \Rightarrow x_1 = k$$

$\begin{pmatrix} 1 \\ 1 \\ 1 \end{pmatrix}$ is the Eigenvector of $\lambda = 4$.

---

6. $\mathbf{A} - \lambda\mathbf{I} = \begin{pmatrix} 2 & 1 & 1 \\ 1 & 2 & 1 \\ 1 & 1 & 2 \end{pmatrix} - \begin{pmatrix} \lambda & 0 & 0 \\ 0 & \lambda & 0 \\ 0 & 0 & \lambda \end{pmatrix} = \begin{pmatrix} 2-\lambda & 1 & 1 \\ 1 & 2-\lambda & 1 \\ 1 & 1 & 2-\lambda \end{pmatrix}$

$$= (2-\lambda)\begin{vmatrix} 2-\lambda & 1 \\ 1 & 2-\lambda \end{vmatrix} - \begin{vmatrix} 1 & 1 \\ 1 & 2-\lambda \end{vmatrix} + \begin{vmatrix} 1 & 2-\lambda \\ 1 & 1 \end{vmatrix}$$

$$= (2 - \lambda\left[(2-\lambda)^2 - 1\right] - (1-\lambda) + (-1+\lambda) = 0$$

$$(2-\lambda)(1-\lambda)(3-\lambda) - 2 + 2\lambda = 0$$

$$(2-\lambda)(1-\lambda)(3-\lambda) + 2(\lambda - 1) = 0$$

$$(1-\lambda)(-3\lambda - 2\lambda + j^2) = (1-\lambda)(\lambda^2 - 5j + 4)$$
$$= (1-\lambda)(\lambda - 4)(\lambda - 1) = 0$$

$\lambda = 1$ twice $\lambda = 4$

Diagonal matrix $\begin{pmatrix} 1 & 0 & 0 \\ 0 & 1 & 0 \\ 0 & 0 & 4 \end{pmatrix}$.

---

7. The vector equation of the line is $\mathbf{r} = (\mathbf{i} + 2\mathbf{j} - 3\mathbf{k}) + \lambda(-3\mathbf{i} - 5\mathbf{j} + 4\mathbf{k})$.

The position vector is $\mathbf{i} + 2\,\mathbf{j} - 3\,\mathbf{k}$ and the direction vector is $-3\mathbf{i} + 5\mathbf{j} - -4\mathbf{k}$

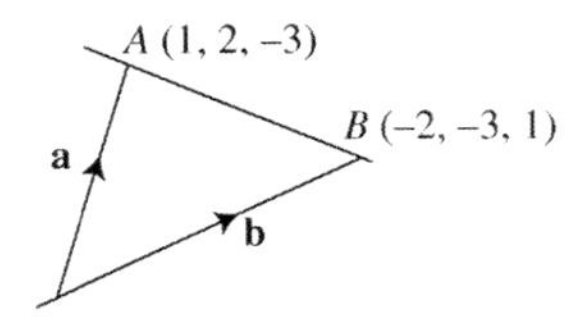

$$d = \frac{|\mathbf{b} \times (\mathbf{p} - \mathbf{a})|}{|\mathbf{b}|} \qquad \text{where } d = \frac{\sqrt{75}}{\sqrt{50}} = 1.22$$

$\mathbf{p} - \mathbf{a} = -\mathbf{i} - 2\mathbf{j} + 3\mathbf{k}$ where $\mathbf{p} = 0$ (origin),

$\mathbf{a} = \mathbf{i} + 2\mathbf{j} - 3\mathbf{k} \qquad \mathbf{b} = -3\mathbf{i} - 5\mathbf{j} + 4\mathbf{k}$

the direction vector $= -2\mathbf{i} - 3\mathbf{j} + \mathbf{k} - (\mathbf{i} - 2\mathbf{j} + 3\mathbf{k}) = -3\mathbf{i} - 5\mathbf{j} + 4\mathbf{k}$

$$|\mathbf{b}| = \sqrt{(-3)^2 + (-5)^2 + 4^2} = \sqrt{9 + 25 + 16} = \sqrt{50}$$

$$\mathbf{b} \times (\mathbf{p} - \mathbf{a}) = (-3\mathbf{i} - 5\mathbf{j} + 4\mathbf{k}) \times (-\mathbf{i} - 2\mathbf{j} + 3\mathbf{k})$$

$$\begin{pmatrix} \mathbf{i} & \mathbf{j} & \mathbf{k} \\ -3 & -5 & 4 \\ -1 & -2 & 3 \end{pmatrix} = \mathbf{i}(-15 + 8) - \mathbf{j}(-9 + 4) + \mathbf{k}(6 - 5) = -7\mathbf{i} + 5\mathbf{j} + \mathbf{k}$$

$$|\mathbf{b} \times (\mathbf{p} - \mathbf{a})| = \sqrt{(-7)^2 + 5^2 + 1^2} = \sqrt{75}.$$

---

8. (a) (i) $y = \operatorname{sech}^2 x$

let $u = \operatorname{sech} x \quad y = u^2 \quad \frac{du}{dx} = -\operatorname{sech} x \tanh x$

$$\frac{dy}{du} = 2u \qquad \frac{dy}{dx} = \frac{dy}{du}\frac{du}{dx} = -2\operatorname{sech}^2 x \tanh x$$

$$\therefore \frac{dy}{dx} = -2\operatorname{sech}^2 x \tanh x.$$

(ii) $y = \sinh^4 x \qquad$ let $u = \sinh x \qquad \frac{dy}{dx} = \cosh x$

$$y = u^4 \qquad \frac{dy}{du} = 4u^3 \qquad \frac{dy}{du} = 4\sinh^3 x \cosh x$$

$$\therefore \frac{dy}{dx} = 4 \sinh^3 x \cosh x.$$

(iii) $y = \sqrt{2}\operatorname{cosech}^{\frac{1}{2}} x \qquad$ let $u = \operatorname{cosech} x$

$$y = \sqrt{2}u^{\frac{1}{2}} \qquad \frac{du}{dx} = -\operatorname{cosech} x \coth x$$

$$\frac{dy}{du} = \frac{1}{2}\sqrt{2}\, u^{-\frac{1}{2}}, \qquad \frac{dy}{dx} = \frac{dy}{du}\frac{du}{dx} - \frac{1}{2}\sqrt{2}\,\frac{\operatorname{cosech} x \coth x}{\operatorname{cosech}^{\frac{1}{2}} x}$$

$$\frac{dy}{dx} = -\frac{1}{2}\sqrt{2}\operatorname{cosech}^{\frac{1}{2}} x \coth x.$$

(b) (i) $y = 3\sinh^{-1}\dfrac{1}{x} \Rightarrow \dfrac{y}{3} = \sinh^{-1}\dfrac{1}{x} \Rightarrow \dfrac{1}{x} = \sinh\dfrac{y}{3}$

$x = \operatorname{cosech}\dfrac{y}{3} \qquad \dfrac{dx}{dy} = -\dfrac{1}{3}\coth\dfrac{y}{3}\operatorname{cosech}\dfrac{y}{3}$

$$\therefore \frac{dy}{dx} = -\frac{3}{\coth\frac{y}{3}\operatorname{cosech}\frac{y}{3}} = -\frac{3}{x\coth\frac{y}{3}} = -\frac{3}{x\sqrt{1+x^2}}$$

where $1 - \coth^2\frac{y}{3} = -\operatorname{cosech}^2\frac{y}{3} \Rightarrow 1 + x^2 = \coth^2\frac{y}{3}$.

(ii) $y = 5\cosh^{-1}(x^2 - 3x + 2) \qquad \dfrac{y}{5} = \cosh^{-1}(x^2 - 3x + 2)$

$x^2 - 3x + 2 = \cosh\dfrac{y}{5}$ differentiating with respect to y we have

$(2x-3)\dfrac{dx}{dy} = \dfrac{1}{5}\sinh\dfrac{y}{5} \Rightarrow \dfrac{dy}{dx} = \dfrac{5(2x-3)}{\sinh\frac{y}{5}}$

$$\therefore \frac{dy}{dx} = \frac{5(2x-3)}{\sqrt{\cosh^2\frac{y}{5} - 1}} = \frac{5(2x-3)}{\sqrt{(x^2-3x+2)^2 - 1}}$$

where $\cosh^2\frac{y}{5} - \sinh^2\frac{y}{5} = 1 \Rightarrow \cosh^2\frac{y}{5} - 1 = \sinh^2\frac{y}{5}$.

Examination Test Papers.

GCE Advanced Level.

Further Pure Mathematics FP3

Test Paper 5

Time: 1 hour and 30 minutes.

Instructions and Information

Candidates may use any calculator allowed by the regulations of their Examination Board.

Full marks are awarded for correct answers to ALL questions.

This paper has eight questions.

You can start working with any question and you must label clearly all parts.

1. (a) Show by using a suitable substitution that $\int \frac{1}{\sqrt{9x^2-16}}\,dx = \frac{1}{3}\cosh^{-1}\frac{3x}{4}$.

(b) Evaluate (i) $\int_2^3 \operatorname{sech}\frac{3}{4}x\,dx$ (ii) $\int_1^2 \tanh\frac{1}{2}x\,dx$. **(10)**

---

2. An ellipse has parametric equations $x = a\cos\theta$, $y = b\sin\theta$, $0 \le \theta \le 2\pi$,

$a > b$, where $b^2 = a^2(1-e^2)$.

Show that the length of the arc, s, of the ellipse from $\theta = 0$ to $\theta = \pi$ is given by

$s = a\int_0^{\pi} \sqrt{1 - e^2\cos^2\theta}\,d\theta$. **(10)**

---

3. If $I_n = \int_0^{\infty} \frac{1}{(1+x^2)^n}\,dx$, where $n \in \mathbb{Z}^+$, show that

$I_{n-1} = \frac{2n-2}{2n-3} I_n$, for $n \ge 2$.

Hence evaluate I_3, I_5, I_7. **(12)**

---

4. Solve the hyperbolic equation

$$6\sinh x - 2\cosh x = 7.$$ **(7)**

---

5. Find the characteristic equation of the matrix $\mathbf{A} = \begin{pmatrix} 3 & 2 & 2 \\ 2 & 2 & 0 \\ 2 & 0 & 4 \end{pmatrix}$ and obtain the

eigenvalues of the matrix. **(8)**

---

6. Find the equation of the plane which contains three points with coordinates

$A(1, 2, 3)$, $B(-2, 3, -1)$, $C(-1, 4, 2)$. **(12)**

---

7. The position vectors of the points A, B and C are $\mathbf{a} = -2\mathbf{i} + 3\mathbf{j} - 5\mathbf{k}$,

$\mathbf{b} = \mathbf{i} + 4\mathbf{j} + 7\mathbf{k}$, $\mathbf{c} = 3\mathbf{i} - 2\mathbf{j} + 2\mathbf{k}$. Determine the volume of the tetrahedron

with the base ΔOBC and vertex A. **(8)**

---

8. Sketch the graph $x^2 = y$.

Find the length of this curve between the limits $x = -3$ and $x = -1$. **(8)**

TOTAL FOR PAPER: 75 MARKS

Examination Test Papers.

GCE Advanced Level.

Test Paper 5 Solutions

Further Pure Mathematics FP3

1. (a) $\int \frac{1}{\sqrt{9x^2-16}} \mathrm{d}x$

let $x = \frac{4}{3}\cosh y \Rightarrow \frac{\mathrm{d}x}{\mathrm{d}y} = \frac{4}{3}\sinh y \quad \text{since} \quad \cosh^2 y - \sinh^2 y = 1$

$$\int \frac{1}{\sqrt{9x^2-16}}\,\mathrm{d}x = \int \frac{1}{\sqrt{9\left(\frac{4}{3}\cosh y\right)^2 - 16}}\,\frac{4}{3}\sinh y\,\mathrm{d}y$$

$$= \int \frac{4}{3}\,\frac{\sinh y}{4\sqrt{\cosh^2 y - 1}}\,\mathrm{d}y = \frac{1}{3}y$$

$$= \frac{1}{3}\cosh^{-1}\frac{3x}{4}.$$

(b) (i) $\int_2^3 \operatorname{sech}\frac{3}{4}x\,\mathrm{d}x = \frac{4}{3}\left[\tan^{-1} e^{\frac{3x}{4}}\right]_2^3$

$$= \frac{4}{3}\tan^{-1} e^{2.25} - \frac{4}{3}\tan^{-1} e^{\frac{3}{2}} = 1.954 - 1.8017 = 0.152$$

(ii) $\int_1^2 \tanh\frac{1}{2}x\,\mathrm{d}x = \left[2\ln\cosh\frac{1}{2}x\right]_1^2$

$$= 2\ln\cosh 1 - 2\ln\cosh\frac{1}{2}$$

$$= 0.86756 - 0.24022 = 0.627.$$

---

2. Length of arc, $s = \int_\alpha^\beta \sqrt{\left(1 + \frac{\mathrm{d}y}{\mathrm{d}x}\right)^2}\,\mathrm{d}x$

$x = a\cos\theta \qquad \frac{\mathrm{d}x}{\mathrm{d}\theta} = -a\sin\theta \qquad y = b\sin\theta \qquad \frac{\mathrm{d}y}{\mathrm{d}\theta} = b\cos\theta$

$$\frac{\frac{\mathrm{d}y}{\mathrm{d}\theta}}{\frac{\mathrm{d}x}{\mathrm{d}\theta}} = \frac{b\cos\theta}{-a\sin\theta} = -\frac{b}{a}\cot\theta$$

$$\int \sqrt{1+\left(\frac{\mathrm{d}y}{\mathrm{d}x}\right)^2}\,\mathrm{d}x = \int_0^{\pi} \sqrt{1+\left(-\frac{b}{a}\right)^2 \cot^2\theta}\,(-a\sin\theta)\,\mathrm{d}\theta$$

$$= \int_0^{\pi} \sqrt{\frac{a^2+b^2\cot^2\theta}{a^2}}\,(-a\sin\theta)\,\mathrm{d}\theta$$

$$= \int_0^{\pi} \sqrt{\frac{a^2+a^2\left(1-e^2\right)\cot^2\theta}{a^2}}\,(-a\sin\theta)\,\mathrm{d}\theta$$

$$= \int_0^{\pi} \sqrt{1+\cot^2\theta - e^2\cot^2\theta}\,(-a\sin\theta)\,\mathrm{d}\theta$$

$$= -a\int_0^{\pi} \sqrt{(\operatorname{cosec}^2\theta - e^2\cot^2\theta)\sin^2\theta}\,\mathrm{d}\theta$$

$$= a\int_0^{\pi} \sqrt{1-e^2\cos^2\theta}\,\mathrm{d}\theta.$$

---

3. $I_n = \int_0^{\infty} \frac{1}{\left(1+x^2\right)^n}\,\mathrm{d}x \qquad I_{n-1} = \int_0^{\infty} \frac{1}{\left(1+x^2\right)^{n-1}}\,\mathrm{d}x$

$$I_{n-1} = \int_0^{\infty} \underset{②}{\left(1+x^2\right)^{-n+1}}\,\underset{①}{\mathrm{d}x} = \left[x\left(1+x^2\right)^{-n+1}\right]_0^{\infty} - \int_0^{\infty} x\,(-n+1)\left(1+x^2\right)^{-n}(2x)\,\mathrm{d}x$$

$$= 2(n-1)\int_0^{\infty} x^2\left(1+x^2\right)^{-n}\mathrm{d}x = 2(n-1)\int_0^{\infty}\left(x^2+1-1\right)\left(1+x^2\right)^{-n}\mathrm{d}x$$

$$= 2(n-1)\int_0^{\infty}\left[\left(1+x^2\right)^{-n+1} - \left(1+x^2\right)^{-n}\right]\mathrm{d}x$$

$$I_{n-1} = 2(n-1)\int_0^{\infty}\frac{1}{\left(1+x^2\right)^{n-1}}\,\mathrm{d}x - 2(n-1)\int_0^{\infty}\frac{1}{\left(1+x^2\right)^{n}}\,\mathrm{d}x$$

$$I_{n-1} = 2(n-1)\,I_{n-1} - 2(n-1)\,I_n$$

$$I_{n-1}(1-2n+2) = -2(n-1)\,I_n$$

$$\therefore \boxed{I_{n-1} = \frac{2n-2}{2n-3}\,I_n}$$

$$I_n = \frac{2n-3}{2n-2}I_{n-1}$$

$$n = 3, \quad I_3 = \frac{3}{4} I_2, \quad I_2 = \frac{1}{2} I_1$$

$$I_1 = \int_0^{\infty} \frac{1}{1+x^2}\,dx = \left[\tan^{-1} x\right]_0^{\infty} = \frac{\pi}{2} \qquad I_3 = \frac{3}{4} \times \frac{1}{2} \times \frac{\pi}{2} = \frac{3\pi}{16}$$

$n = 5$

$$I_5 = \frac{7}{8} I_4 = \frac{7}{8}\frac{5}{6} I_3 = \frac{7}{8}\frac{5}{6}\frac{3\pi}{16} = \frac{105}{768}\pi = \frac{35}{256}\pi$$

$n = 7$

$$I_7 = \frac{11}{12} I_6 = \frac{11}{12}\frac{9}{10} I_5$$

$$I_7 = \frac{11}{12}\frac{9}{10}\frac{105}{768}\pi = \frac{10395}{92160}\pi = \frac{231}{2048}\pi.$$

---

4. $6\sinh x - 2\cosh x = 7$

$$\sinh x = \frac{e^x - e^{-x}}{2} \text{ and } \cosh x = \frac{e^x + e^{-x}}{2}$$

$$6\,\frac{e^x - e^{-x}}{2} - 2\,\frac{e^x + e^{-x}}{2} = 7 \qquad 3e^x - 3e^{-x} - e^x - e^{-x} - 7 = 0$$

$$2e^x - 4e^{-x} - 7 = 0$$

multiplying each side by e^x

$$2e^{2x} - 7e^x - 4 = 0$$

$$e^x = \frac{7 \pm \sqrt{49+32}}{4} = \frac{7 \pm 9}{4}$$

$e^x = 4$ or $e^x = -\frac{1}{2}$ which is disregarded

$$\boxed{x = \ln 4}$$

---

5. $f(\lambda) = |\mathbf{A} - \lambda\mathbf{I}| = \begin{vmatrix} 3-\lambda & 2 & 2 \\ 2 & 2-\lambda & 0 \\ 2 & 0 & 4-\lambda \end{vmatrix} = 0$

$$f(\lambda) = (3-\lambda)\begin{vmatrix} 2-\lambda & 0 \\ 0 & 4-\lambda \end{vmatrix} - 2\begin{vmatrix} 2 & 0 \\ 2 & 4-\lambda \end{vmatrix} + 2\begin{vmatrix} 2 & 2-\lambda \\ 2 & 0 \end{vmatrix}$$

$$= (3-\lambda)(2-\lambda)(4-\lambda) - 2(8-2\lambda) + 2(-4+2\lambda)$$

$$= (3-\lambda)(2-\lambda)(4-\lambda) - 4(4-\lambda) - 4(2-\lambda)$$

$$= \left(6 - 2\lambda - 3\lambda + \lambda^2\right)(4-\lambda) - 16 + 4\lambda - 8 + 4\lambda$$

$$= 24 - 8\lambda - 12\lambda + 4\lambda^2 - 6\lambda + 2\lambda^2 + 3\lambda^2 - \lambda^3 - 16 + 4\lambda - 8 + 4\lambda$$

$$= -\lambda^3 + 9\lambda^2 - 18\lambda = -\lambda\left(\lambda^2 - 9\lambda + 18\right) = 0$$

$$\lambda = 0 \qquad \lambda = \frac{9 \pm \sqrt{81-72}}{2} = \frac{9+3}{2}$$

$$\boxed{\lambda = 0} \quad \text{or} \quad \boxed{\lambda = 3} \quad \text{or} \quad \boxed{\lambda = 6}.$$

---

6. The equation of the plane is $ax + by + cz = \mathrm{d}$,

where $a : b : \mathrm{c}$ are the direction ratios of the line which is perpendicular to the plane.

Since $A(1, 2, 3)$ lies on the plane when this will satisfy the

equation $\quad a + 2b + 3c = \mathrm{d} \quad \ldots(1)$

since $B(-2, 3, -1)$ lies also on the plane then this will also satisfy the

equation $-2a + 3b - c = \mathrm{d} \quad \ldots(2)$

and finally $C(-1, 4, 2)$ lies on the plane then $-a + 4b + 2c = \mathrm{d} \quad \ldots(3)$

Solving equations (1), (2) and (3) in terms of d

$$a + 2b + 3c = \mathrm{d} \quad \ldots(1)$$

$$-2a + 3b - c = \mathrm{d} \quad \ldots(2)$$

$$-a + 4b + 2c = \mathrm{d} \quad \ldots(3)$$

adding (1) and (3) $6b + 5c = 2\mathrm{d} \quad \ldots(4)$

$(1) \times 2 \qquad 2a + 4b + 6c = 2\text{d}$

$(2) \qquad -2a + 3b - c = \text{d}$

$7b + 5c = 3\text{d} \qquad \ldots(5)$

$(5) - (4) \qquad \boxed{b = \text{d}}$

substituting these values in (4)

$6d + 5c = 2\text{d} \qquad 5c = -4\text{d} \qquad \boxed{c = -\frac{4}{5}\text{d}}$

From (1) $\quad a + 2b + 3c = \text{d}$

$$a + 2d + 3\left(-\frac{4}{5}\text{d}\right) = \text{d} \Rightarrow a = \frac{7}{5}\text{d}$$

$a = \frac{7}{5}\text{d}, \qquad b = \text{d}, \qquad c = -\frac{4}{5}\text{d}$

$\therefore$ The equation of the plane is given $\qquad 7x + 5y - 4 = 5.$

If a plane contains the origin $O(0, 0, 0)$, $\text{d} = 0$.

7.

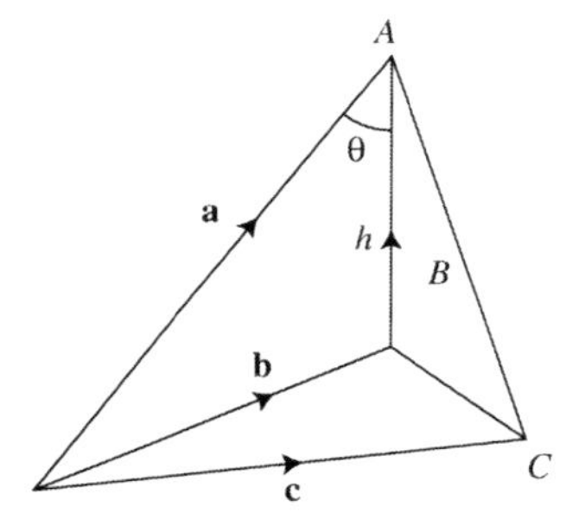

Referring to a fixed point 0,

the position vectors of the point A, B and C are

a, **b** and **c** respectively.

$$V = \frac{1}{6}\left|(\mathbf{b} \times \mathbf{c}) . \mathbf{a}\right|$$

$$\mathbf{b} \times \mathbf{c} = \begin{vmatrix} \mathbf{i} & \mathbf{j} & \mathbf{k} \\ 1 & 4 & 7 \\ 3 & -2 & 2 \end{vmatrix} = \mathbf{i}(8 + 14) - \mathbf{j}(2 - 21) + \mathbf{k}(-2 - 12) = 22\mathbf{i} + 19\mathbf{j} - 14\mathbf{k}$$

$$V = \tfrac{1}{6}\ |(22\mathbf{i} + 19\mathbf{j} - 14\mathbf{k}) . (-2\mathbf{i} + 3\mathbf{j} - 5\mathbf{k})|$$

$$V = \tfrac{1}{6}\ |-44 + 57 + 70| = \tfrac{1}{6}(83) = 13.8 \text{ cubic units.}$$

8.

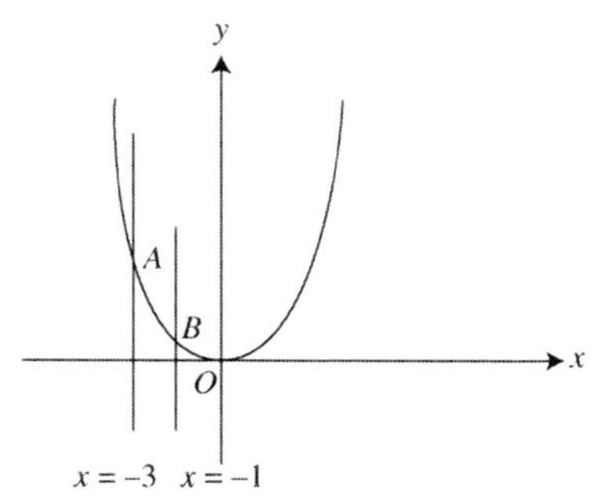

$$\text{Length of arc } AB = \int_{-3}^{-1} \sqrt{1 + \left(\frac{dy}{dx}\right)^2}\, dx$$

$$= \int_{-3}^{-1} \sqrt{1 + 4x^2}\, dx$$

$x^2 = y \qquad \text{where} \quad \dfrac{dy}{dx} = 2x$

let $2x = \sinh y \qquad 2\dfrac{dx}{dy} = \cosh y$

$$\text{Length of arc } AB = \int_{-2.49}^{-1.44} \sqrt{1 + \sinh^2 y}\, \frac{\cosh y}{2}\, dy = \frac{1}{2}\int_{-2.49}^{-1.44} \cosh^2 y\, dy$$

$$= \frac{1}{2}\int_{-2.49}^{-1.44} \frac{2y+1}{2}\, dy$$

$$= \frac{1}{4}\int_{-2.49}^{-1.44} (\cosh 2y + 1)\, dy = \frac{1}{4}\left(\frac{\sinh 2y + y}{2}\right)_{-2.49}^{-1.44}$$

$$= \frac{1}{4}(-4.44 - 1.44 + 36.4 + 2.49) = 8.25 \text{ units.}$$

Examination Test Papers.

GCE Advanced Level.

Further Pure Mathematics FP3

Test Paper 6

Time: 1 hour and 30 minutes.

Instructions and Information

Candidates may use any calculator allowed by the regulations of their Examination Board.

Full marks are awarded for correct answers to ALL questions.

This paper has eight questions.

You can start working with any question and you must label clearly all parts.

1. Determine the following indefinite integrals:

(i) $\displaystyle\int \cosh^2 x\, \mathrm{d}x$ (ii) $\displaystyle\int \sinh^2 x\, \mathrm{d}x$ (iii) $\displaystyle\int \tanh^2 x\, \mathrm{d}x$

(iv) $\displaystyle\int \coth^2 x\, \mathrm{d}x$ (v) $\displaystyle\int \operatorname{sech}^3 x\, \mathrm{d}x$ (vi) $\displaystyle\int \operatorname{cosech}^2 x\, \mathrm{d}x$ **(12)**

---

2. (a) Show that (i) $\dfrac{\mathrm{d}}{\mathrm{d}x}\left(\cosh^{-1} x\right) = \dfrac{1}{\sqrt{x^2 - 1}}$

(ii) $\dfrac{\mathrm{d}}{\mathrm{d}x}\left(\sinh^{-1} x\right) = \dfrac{1}{\sqrt{1 + x^2}}.$

(b) If $y = \dfrac{\cosh^{-1} x}{\sinh^{-1} x}$ determine $\dfrac{\mathrm{d}y}{\mathrm{d}x}$. **(8)**

---

3. Show that a reduction formula for $I_n = \displaystyle\int \sec^n x\, \mathrm{d}x$

is given by the formula

$I_n = \dfrac{1}{n-1} \tan x \sec^{n-2} x + \dfrac{n-2}{n-1} I_n{-2}$ where $n \geq 2$,

and hence find I_5. **(10)**

---

4. The line $y = mx + c$ is tangent to the ellipse $\dfrac{x^2}{4} + \dfrac{y^2}{9} = 1$.

Find the relationship between the values of m and c. **(8)**

---

5. If $\mathbf{A} = \begin{pmatrix} 2 & -2 & -1 \\ 1 & 2 & -2 \\ 2 & 1 & 2 \end{pmatrix}$ find $\mathbf{A}^{-1}$ and $\mathbf{A}^{\mathrm{T}}$ and show that $\mathbf{A}$, $\mathbf{A}^{-1}$ and $\mathbf{A}^{\mathrm{T}}$

are orthogonal matrices. **(12)**

---

6. Find the equation of the plane which contains the two lines

$$l_1 : \frac{x+2}{4} = \frac{y+6}{3} = \frac{z-2}{2}$$

$$l_2 : \frac{x-3}{1} = \frac{y+1}{2} = \frac{z-7}{3}.$$ **(10)**

---

7. Find the volume of a parallelepiped where O is the origin and A, B and C are the points $(-3, 2, 4)$, $(2, -3, -1)$ and $(3, -1, 2)$ respectively. **(9)**

---

8. Integrate by parts:-

(i) $\int \coth^{-1} x \, dx$ (ii) $\int \text{sech}^{-1} dx$ (iii) $\int \text{cosech}^{-1} x \, dx$. **(6)**

TOTAL FOR PAPER: 75 MARKS

Examination Test Papers.

GCE Advanced Level.

Test Paper 6 Solutions

Further Pure Mathematics FP3

1. (i) $\int \cosh^2 x \, dx = \int \frac{\cosh 2x + 1}{2} dx = \frac{\sinh 2x}{4} + \frac{1}{2}x + c$

where $\cosh 2x = 2\cosh^2 x - 1 \Rightarrow \cosh^2 x = \frac{\cosh 2x + 1}{2}$.

(ii) $\int \sinh^2 x \, dx = \int \frac{\cosh 2x - 1}{2} dx = \frac{\sinh 2x}{4} - \frac{1}{2}x + c$

where $\cosh 2x = 1 + 2\sinh^2 x \Rightarrow \sinh^2 x = \frac{\cosh 2x - 1}{2}$.

(iii) $\int \tanh^2 x \, dx = \int \left(1 - \operatorname{sech}^2 x\right) dx = x - \tanh + c$

where $1 - \tanh^2 x = \operatorname{sech}^2 x \Rightarrow \tanh^2 x = 1 - \operatorname{sech}^2 x$.

(iv) $\int \coth^2 x \, dx = \int \left(1 + \operatorname{cosech}^2 x\right) dx = x - \coth x + c$

where $1 - \coth^2 x = -\operatorname{cosech}^2 x \Rightarrow \coth^2 x = 1 + \operatorname{cosech}^2 x$.

(v) $\int \operatorname{sech}^2 x \, dx = \tanh x + c$.

(vi) $\int \operatorname{cosech}^2 x \, dx = -\coth x + c$.

---

2. (a) (i) $\frac{d}{dx}\left(\cosh^{-1} x\right) = \frac{1}{\sqrt{x^2 - 1}}$

let $y = \cosh^{-1} x$

$x = \cosh y$

$$\frac{dx}{dy} = \sinh y \Rightarrow \frac{dy}{dx} = \frac{1}{\sinh y} = \frac{1}{\sqrt{\cosh^2 y - 1}} = \frac{1}{\sqrt{x^2 - 1}}.$$

(ii) $\frac{d}{dx}\left(\sinh^{-1} x\right) = \frac{1}{\sqrt{1 + x^2}}$.

Let $y = \sinh^{-1} x$

$x = \sinh y$

$$\frac{dx}{dy} = \cosh y \Rightarrow \frac{dy}{dx} = \frac{1}{\cosh y} = \frac{1}{\sqrt{1 + \sinh^2 y}} = \frac{1}{\sqrt{x^2 + 1}}.$$

(b) $y = \dfrac{\cosh^{-1} x}{\sinh^{-1} x}$

$$\frac{dy}{dx} = \left[\frac{1}{\sqrt{x^2-1}}\sinh^{-1} x - \cosh^{-1} x \frac{1}{\sqrt{x^2+1}}\right] \times \frac{1}{\left[\sinh^{-1} x\right]^2}.$$

---

3. $I_n = \displaystyle\int \sec^n x \, dx = \int \underset{②}{\sec^{n-2} x} \, \underset{①}{\sec^2 x} \, dx = \tan x \sec^{n-2} x$

$$- \int \tan x \, (n-2) \sec^{n-3} x \sec x \tan x \, dx$$

$$= \tan x \sec^{n-2} x - (n-2) \textstyle\int \tan^2 x \sec^{n-2} x \, dx$$

but $\displaystyle\int \tan^2 x \sec^{n-2} x \, dx = \int \left(\sec^2 x - 1\right) \sec^{n-2} x \, dx$

$$I_n = \tan x \sec^{n-2} x - (n-2) \int \sec^n x \, dx + (n-2) \int \sec^{n-2} x \, dx$$

$$I_n \, (1 + n - 2) = \tan x \sec^{n-2} x + (n-2) \, I_{n-2}$$

$$\boxed{I_n = \int \sec^n x \, dx = \frac{1}{n-1} \tan x \sec^{n-2} x + \frac{n-2}{n-1} I_{n-2}}$$

$n = 5 \qquad I_5 = \dfrac{1}{4} \tan x \sec^3 x + \dfrac{3}{4} I_3 \qquad I_3 = \dfrac{1}{2} \tan x \sec x + \dfrac{1}{2} I_1$

$$I_1 = \int \sec x \, dx = \ln |\sec x + \tan x|$$

$$\therefore I_5 = \frac{1}{4} \tan x \sec^3 x + \frac{3}{4} \left(\frac{1}{2} \tan x \sec x + \frac{1}{2} I_1\right)$$

$$= \frac{1}{4} \tan x \sec^3 x + \frac{3}{8} \tan x \sec x + \frac{3}{8} \ln |\sec x + \tan x| \, .$$

4.
$$\frac{x^2}{4} + \frac{y^2}{9} = 1.$$

If $y = 0$, $x = \pm 2$, if $x = 0$, $y = \pm 3$.

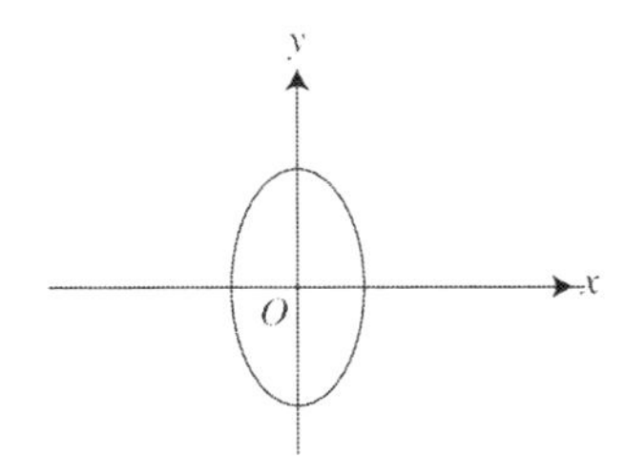

$9x^2 + 4y^2 = 36 \qquad 9x^2 + 4(mx + c)^2 = 36$

$9x^2 + 4m^2x^2 + 8mc\,x + 4c^2 - 36 = 0 \qquad x^2(9 + 4m^2) + 8mc\,x + 4c^2 - 36 = 0.$

For the line to be tangent the discriminant

$D = b^2 - 4ac = 64m^2c^2 - 4(9 + 4m^2) \times (4c^2 - 36) = 0$

$64m^2c^2 = 4(36c^2 + 16m^2c^2 - 324 - 144m^2)$

$64m^2c^2 = 144c^2 + 64m^2c^2 - 1296 - 576m^2$

$144c^2 - 576m^2 = 1296 \qquad 36c^2 - 144m^2 = 324$

$$\boxed{9c^2 - 36m^2 = 81}$$

---

5. $\mathbf{A} = \begin{pmatrix} 2 & -2 & -1 \\ 1 & 2 & -2 \\ 2 & 1 & 2 \end{pmatrix}$

$$\mathbf{A}^* = \begin{pmatrix} 6 & -6 & -3 \\ 3 & 6 & -6 \\ 6 & 3 & 6 \end{pmatrix} \qquad \mathbf{A}^{*\mathrm{T}} = \begin{pmatrix} 6 & 3 & 6 \\ -6 & 6 & 3 \\ -3 & -6 & 6 \end{pmatrix}$$

$|\mathbf{A}| = 2 \times 6 + 2 \times 6 - 1(-3) = 27$

$$\mathbf{A}^{-1} = \frac{\mathbf{A}^{*\mathrm{T}}}{|\mathbf{A}|} = \frac{1}{27}\begin{pmatrix} 6 & 3 & 6 \\ -6 & 6 & 3 \\ -3 & -6 & 6 \end{pmatrix} = \frac{1}{9}\begin{pmatrix} 2 & 1 & 2 \\ -2 & 2 & 1 \\ -1 & -2 & 2 \end{pmatrix}$$

$$\mathbf{A}^{\mathrm{T}} = \begin{pmatrix} 2 & 1 & 2 \\ -2 & 2 & 1 \\ -1 & -2 & 2 \end{pmatrix}$$

$$\mathbf{A}^{-1}\mathbf{A} = \frac{1}{9}\begin{pmatrix} 2 & 1 & 2 \\ -2 & 2 & 1 \\ -1 & -2 & 2 \end{pmatrix}\begin{pmatrix} 2 & -2 & -1 \\ 1 & 2 & -2 \\ 2 & 1 & 2 \end{pmatrix} = \frac{1}{9}\begin{pmatrix} 4+1+4 & 0 & 0 \\ 0 & 4+4+1 & 0 \\ 0 & 0 & 1+4+4 \end{pmatrix}$$

$$= \begin{pmatrix} 1 & 0 & 0 \\ 0 & 1 & 0 \\ 0 & 0 & 1 \end{pmatrix} = \mathbf{I}$$

$$\mathbf{A}^{\mathrm{T}}\mathbf{A} = \begin{pmatrix} 2 & 1 & 2 \\ -2 & 2 & 1 \\ -1 & -2 & 2 \end{pmatrix}\begin{pmatrix} 2 & -2 & -1 \\ 1 & 2 & -2 \\ 2 & 1 & 2 \end{pmatrix} = \begin{pmatrix} 9 & 0 & 0 \\ 0 & 9 & 0 \\ 0 & 0 & 9 \end{pmatrix} = 9\mathbf{I}.$$

Orthogonal vectors are vectors which are at right angles to each other.

---

6. Let l_3 be a line perpendicular to both lines l_1 and l_2, with direction ration $a : b : c$.

$$l_1: \frac{x+2}{4} = \frac{y+6}{3} = \frac{z-2}{2}$$

$$l_2: \frac{x-3}{1} = \frac{y+1}{2} = \frac{z-7}{3}$$

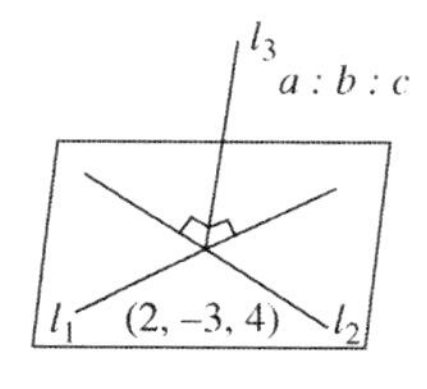

Equation of a plane containing two lines.

Since l_3 is perpendicular to l_1 then $\quad 4a + 3b + 2c = 0 \quad \ldots(1)$

Since l_3 is perpendicular to l_2 then $\quad a + 2b + 3c = 0 \quad \ldots(2)$

$$\begin{aligned} 4a + 3b + 2c &= 0 \\ -4a - 8b - 12c &= 0 \qquad (2) \times -4 \\ \hline -5b - 10c &= 0 \end{aligned}$$

$\frac{b}{c} = \frac{-2}{1}$ From (2) $\quad a - 4c + 3c = 0,\ \boxed{a = c}$

$\frac{a}{c} = \frac{1}{1} \quad \frac{\frac{a}{c}}{\frac{b}{c}} = -\frac{1}{2} \Rightarrow \frac{a}{b} = -\frac{1}{2}$

$a : b : c = 1 : -2 : 1$

$x - 2y + z = 1 \times 2 + (-2)(-3) + 1x4 = 12$

$$\boxed{x - 2y + z = 12}$$

7.

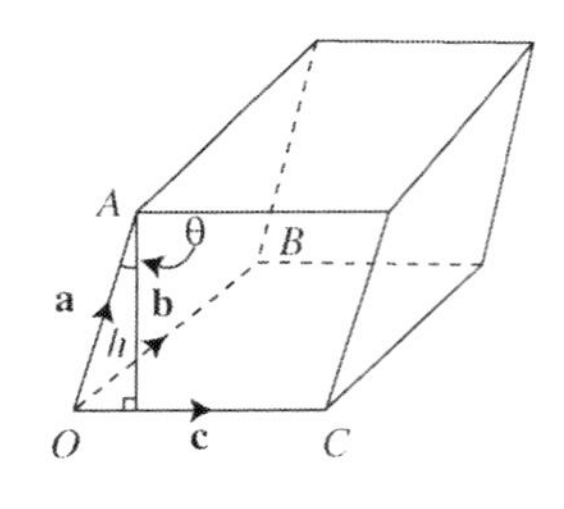

$$\cos\theta = \frac{h}{\mathbf{a}}$$

$$V = |\mathbf{b}\times\mathbf{c}|\,h = |\mathbf{b}\times\mathbf{c}|\,|\mathbf{a}|\cos\theta$$

$$V = |(\mathbf{b}\times\mathbf{c})\,.\,\mathbf{a}|$$

$\mathbf{a} = -3\mathbf{i}+2\mathbf{j}+4\mathbf{k} \qquad \mathbf{b} = 2\mathbf{i}-3\mathbf{j}-\mathbf{k} \qquad \mathbf{c} = 3\mathbf{i}-\mathbf{j}+2\mathbf{k}$

$V = |(\mathbf{b}\times\mathbf{c})\,.\,\mathbf{a}|$

$$\mathbf{b}\times\mathbf{c} = \begin{vmatrix} \mathbf{i} & \mathbf{j} & \mathbf{k} \\ 2 & -3 & -1 \\ 3 & -1 & 2 \end{vmatrix} = \mathbf{i}(-6-1) - \mathbf{j}(4+3) + \mathbf{k}(-2+9)$$
$$= -7\mathbf{i} - 7\mathbf{j} + 7\mathbf{k}$$

$(\mathbf{b}\times\mathbf{c})\,.\,\mathbf{a} = (-7\mathbf{i}-7\mathbf{j}+7\mathbf{k})\,.\,(-3\mathbf{i}+2\mathbf{j}+4\mathbf{k}) = 21 - 14 + 28 = 35$

$V = |(\mathbf{b}\times\mathbf{c})\,.\,\mathbf{a}| = 35$ cubic units.

Alternatively

$V = |(\mathbf{a}\times\mathbf{b})\,.\,\mathbf{c}|$

$$\mathbf{a}\times\mathbf{b} = \begin{vmatrix} \mathbf{i} & \mathbf{j} & \mathbf{k} \\ -3 & 2 & 4 \\ 2 & -3 & -1 \end{vmatrix} = \mathbf{i}(-2+12) - \mathbf{j}(3-8) + \mathbf{k}(9-4)$$
$$= 10\mathbf{i} + 5\mathbf{j} + 5\mathbf{k}$$

$(\mathbf{a}\times\mathbf{b})\,.\,\mathbf{c} = (10\mathbf{i}+5\mathbf{j}+5\mathbf{k})\,.\,(3\mathbf{i}-\mathbf{j}+2\mathbf{k})$

$\qquad = 30 - 5 + 10 = 35$ cubic units

$\therefore\ (\mathbf{b}\times\mathbf{c})\,.\,\mathbf{a} = (\mathbf{a}\times\mathbf{b})\,.\,\mathbf{c}.$

---

8. (i) $$\int \underset{②}{\coth^{-1} x}\, \underset{①}{dx} = x\coth^{-1}x - \int x\frac{1}{1-x^2}\,dx$$

$$= x\coth^{-1}x + \frac{1}{2}\int \frac{d\left(1-x^2\right)}{1-x^2}$$

$$= x\coth^{-1}x + \frac{1}{2}\ln\left|1-x^2\right| + c.$$

(ii) $\displaystyle\int \underset{②}{\operatorname{sech}^{-1}x}\, \underset{①}{dx} = x \operatorname{sech}^{-1}x - \int -x \frac{1}{x\left(1-x^2\right)^{\frac{1}{2}}}\, dx$

$$= x \operatorname{sech}^{-1}x + \int \frac{1}{\left(1-x^2\right)^{\frac{1}{2}}}\, dx$$

$$= x \operatorname{sech}^{-1}x + \sin^{-1}x + c.$$

(iii) $\displaystyle\int \underset{②}{\operatorname{cosech} x}\, \underset{①}{dx} = x \operatorname{cosech}^{-1}x - \int x \left(-\frac{1}{x\left(1+x^2\right)^{\frac{1}{2}}}\right) dx$

$$= x \operatorname{cosech}^{-1}x + \int \frac{1}{\left(1-x^2\right)^{\frac{1}{2}}}\, dx$$

$$= x \operatorname{cosech}^{-1}x + \sinh^{-1}x + c.$$

Examination Test Papers.

GCE Advanced Level.

Further Pure Mathematics FP3

Test Paper 7

Time: 1 hour and 30 minutes.

Instructions and Information

Candidates may use any calculator allowed by the regulations of their Examination Board.

Full marks are awarded for correct answers to ALL questions.

This paper has eight questions.

You can start working with any question and you must label clearly all parts.

1. Determine the indefinite integrals

(i) $\int \sinh^3 x \, dx$ (ii) $\int \cosh^5 x \, dx$ (iii) $\int \frac{\cosh x}{\sinh^3 x} dx$ (iv) $\int \frac{\sinh x}{\sqrt{\cosh x}} dx$. **(10)**

---

2. If $y = \sqrt{\frac{\cosh 2x + 1}{\cosh 2x - 1}}$ show that

$\frac{dy}{dx} = \frac{1}{2} \ln \left| \frac{\cosh 2x + 1}{\cosh 2x - 1} \right| \left(\frac{\cosh 2x + 1}{\cosh 2x - 1} \right)^{\frac{1}{2}}$. **(6)**

---

3. If $I_n = \int \cot^n x \, dx$, prove that $I_n = -\frac{1}{n-1} \cot^{n-1} x - I_{n-2}$,

where $n \geq 2$ and hence find $\int \cot^5 x \, dx$. **(8)**

---

4. Determine the indefinite integrals:

(i) $\int 2 \sinh 2x \cosh 3x \, dx$ (ii) $\int \cosh 3x \cosh 4x \, dx$.

Justifying odd and even functions and applying Osborne's rule. **(12)**

---

5. Determine the matrix **M** such that the point $P(x, y, z)$ is rotated 90° anticlockwise about oz to the point $P'(x', y', z')$ by the transformation

$$\mathbf{M} \begin{pmatrix} x \\ y \\ z \end{pmatrix} = \begin{pmatrix} x' \\ y' \\ z' \end{pmatrix}.$$ **(8)**

---

6. Find the vector equation of the plane containing the points $A(2, 1, 0)$, $B(3, -1, 1)$ and $C(0, -2, -1)$. **(10)**

---

7. Find the perpendicular distance of a point $C(1, 2, 3)$ from the line with a vector equation $\mathbf{r} = (2\mathbf{i} + 3\mathbf{j} + 4\mathbf{k}) + \mathbf{t}(-3\mathbf{i} + 4\mathbf{j} - \mathbf{k})$. **(9)**

---

8. Integrate by parts:-

(i) $\sinh^{-1} x$ (ii) $\cosh^{-1} x$ (iii) $\tanh^{-1} x$. **(12)**

TOTAL FOR PAPER: 75 MARKS

Examination Test Papers.

GCE Advanced Level.

Test Paper 7 Solutions

Further Pure Mathematics FP3

1. (i) $\int \sinh^3 x \, dx = \int \sinh x \sinh^2 x \, dx = \int \sinh^2 x \, d(\cosh x)$

$$= \int \left(\cosh^2 x - 1\right) d(\cosh x) = \frac{\cosh^3 x}{3} - \cosh x + c$$

(ii) $\int \cosh^5 x \, dx = \int \cosh x \cosh^4 x \, dx = \int \cosh^4 x \, d(\sinh x)$

$\cosh^4 x = \left(1 + \sinh^2 x\right)^2 = 1 + 2\sinh^2 x + \sinh^4 x$

$\therefore \int \cosh^4 x \, d(\sinh x) = \int \left(1 + 2\sinh^2 x + \sinh^4 x\right) d(\sinh x)$

$= \sinh x + \frac{2}{3}\sinh^3 x + \frac{1}{5}\sinh^5 x + c$

(iii) $\int \frac{\cosh x}{\sinh^3 x} dx = \int \frac{d(\sinh x)}{\sinh^3 x} = \int (\sinh x)^{-3} \, d(\sinh x)$

$$= \frac{(\sinh x)^{-2}}{-2} = -\frac{1}{2\sinh^2 x} + c = -\frac{1}{2}\operatorname{cosech}^2 x + c$$

(iv) $\int \frac{\sinh x}{\sqrt{\cosh x}} dx = \int \frac{d(\cosh x)}{(\cosh x)^{\frac{1}{2}}} = \int (\cosh x)^{-\frac{1}{2}} \, d(\cosh x)$

$$= \frac{(\cosh x)^{\frac{1}{2}}}{\frac{1}{2}} + c = 2\sqrt{\cosh x} + c.$$

---

2. $y = \sqrt{\frac{\cosh 2x + 1}{\cosh 2x - 1}} \qquad \ldots (1)$

taking logarithms to the base e of (1)

$\ln y = \ln\sqrt{\frac{\cosh 2x + 1}{\cosh 2x - 1}}$

$\frac{1}{y}\frac{dy}{dx} = \frac{1}{2}\ln|\cosh 2x + 1| - \frac{1}{2}\ln|\cos 2x - 1|$

$\frac{dy}{dx} = \frac{1}{2}\ln\left|\frac{\cosh 2x + 1}{\cosh 2x - 1}\right| \left(\frac{\cosh 2x + 1}{\cosh 2x - 1}\right)^{\frac{1}{2}}.$

---

3. $I_n = \cot^n x \, dx = \int \cot^{n-2} x \cot^2 x \, dx = \int \cot^{n-2} x \left(\operatorname{cosec}^2 x - 1 \right) dx$

$$= \int \cot^{n-2} x \operatorname{cosec}^2 x \, dx - I_{n-2} = -\int \cot^{n-2} x \, d(\cot x) - I_{n-2}$$

$$I_n = \int \cot^n x \, dx = -\frac{\cot^{n-1} x}{n-1} - I_{n-2}$$

$$I_5 = \frac{-\cot^4 x}{4} - I_3 \qquad I_3 = \frac{-\cot^2 x}{2} - I_1$$

$$I_1 = \int \cot x \, dx = \int \frac{\cos x}{\sin x} dx = \int \frac{d(\sin x)}{\sin x} = \ln \sin x$$

$$I_3 = -\frac{\cot^2 x}{2} - \ln \sin x \qquad I_5 = -\frac{\cot^4 x}{4} + \frac{\cot^2 x}{2} + \ln \sin x.$$

---

4. (i) $\int 2 \sinh 2x \cosh 3x \, dx = \int \left[\sinh(2x - 3x) + \sinh(2x + 3x) \right] dx$

$$= \int (-\sinh x + \sinh 5x) \, dx = -\cosh x + \frac{1}{5} \cosh 5x + c$$

since $\sinh(2x - 3x) = \sinh 2x \cosh 3x - \cosh 2x \sinh 3x$

$\sinh(2x + 3x) = \sinh 2x \cosh 3x + \cosh 2x \sinh 3x$

$\sinh(2x - 3x) + \sinh(2x + 3x) = 2 \sinh 2x \cosh 3x$

$\sinh(-x) = \dfrac{e^{-x} - e^{-(-x)}}{2} = \dfrac{e^{-x} - e^{x}}{2} = -\dfrac{e^x - e^{-x}}{2} = -\sinh x$ odd function

(ii) $\int \cosh 3x \cosh 4x \, dx$

but $\cosh(x + y) = \cosh x \cosh y + \sinh x \sinh y$

since $\cos(x + y) = \cos x \cos y - \sin x \sin y$

$\cosh(x + y) = \cosh x \cosh y - i \sinh x \, i \sinh y$

$= \cosh x \cosh y + \sinh x \sinh y$

replacing $\sin x$ by $i \sinh x$ in the circular function identities

$\cosh(x - y) = \cosh x \cosh y - \sinh x \sinh y$

$\cosh(x + y) + \cosh(x - y) = 2\cosh x \cosh y$

$$\therefore \int \cosh 3x \cosh 4x \, dx = \frac{1}{2}\int \left[\cosh(3x + 4x) + \cosh(3x - 4x)\right] dx$$

$$= \int \frac{1}{2}\left[\cosh 7x + \cosh(-x)\right] dx = \frac{\sinh 7x}{14} + \frac{\sinh x}{2} + c$$

$$\cosh(-x) = \frac{e^{-x} + e^{-(-x)}}{2} = \frac{e^{x} + e^{-x}}{2} = \cosh x \text{ an even function.}$$

---

5. The unit vectors, **i**, **j**, **k** can be represented as (1, 0, 0), (0, 1, 0), and (0, 0, 1) respectively or as a column vectors of

$$\begin{pmatrix} 1 \\ 0 \\ 0 \end{pmatrix}, \begin{pmatrix} 0 \\ 1 \\ 0 \end{pmatrix} \text{ and } \begin{pmatrix} 0 \\ 0 \\ 1 \end{pmatrix}.$$

The required matrix **M** is given

$$\mathbf{M} = \begin{pmatrix} 1 & -1 & 0 \\ 1 & 0 & 0 \\ 0 & 0 & 1 \end{pmatrix} \text{ since } \mathbf{M} \text{ maps } \mathbf{i} \to \begin{pmatrix} 0 \\ 1 \\ 0 \end{pmatrix}, \mathbf{j} \to \begin{pmatrix} -1 \\ 0 \\ 0 \end{pmatrix} \text{ and } \mathbf{k} \to \begin{pmatrix} 0 \\ 0 \\ 1 \end{pmatrix},$$

$$\text{so } \begin{pmatrix} 1 \\ 0 \\ 0 \end{pmatrix} \to \begin{pmatrix} 0 \\ 1 \\ 0 \end{pmatrix} \quad \begin{pmatrix} 0 \\ 1 \\ 0 \end{pmatrix} \to \begin{pmatrix} -1 \\ 0 \\ 0 \end{pmatrix} \quad \text{and } \begin{pmatrix} 0 \\ 0 \\ 1 \end{pmatrix} \to \begin{pmatrix} 0 \\ 0 \\ 1 \end{pmatrix},$$

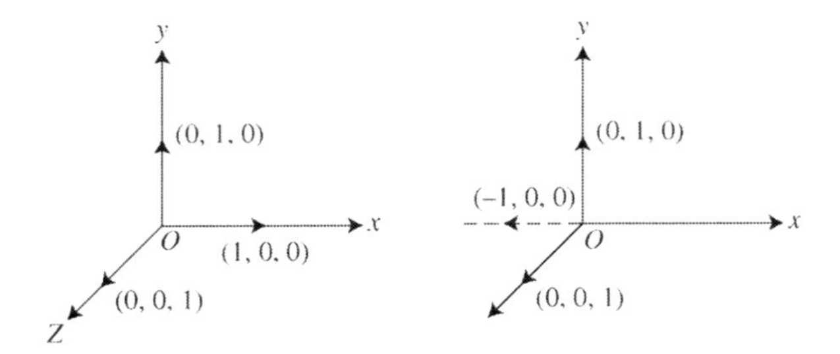

---

6. (a) The parametric equation of this plane is

$$\mathbf{r} = \lambda(2\mathbf{i} + \mathbf{j}) + \mu(3\mathbf{i} - \mathbf{j} + \mathbf{k}) + \nu(-2\mathbf{j} - \mathbf{k})$$

where $\lambda + \mu + \nu = 1 \quad \Rightarrow \quad \nu = 1 - \lambda - \mu$

$$\mathbf{r} = \lambda(2\mathbf{i} + \mathbf{j}) + \mu(3\mathbf{i} - \mathbf{j} + \mathbf{k}) + (1 - \lambda - \mu)(-2\mathbf{j} - \mathbf{k})$$

$$\mathbf{r} = (2\lambda + 3\mu)\mathbf{i} + (\lambda - \mu + 2\lambda + 2\mu - 2)\mathbf{j} + (\mu - 1 + \lambda + \mu)\,\mathbf{k}$$

$$\boxed{\mathbf{r} = (2\lambda + 3\mu)\mathbf{i} + (3\lambda - \mu - 2)\mathbf{j} + (2\mu + \lambda - 1)\mathbf{k}}$$

(b) $x = 2\lambda + 3\mu$, $y = 3\lambda + \mu - 2$, $z = 2\mu + \lambda - 1$.

Eliminating λ and μ from these equations

$$x = 2\lambda + 3\mu, \qquad -3y = -9\lambda - 3\mu + 6$$

$$\boxed{x - 3y = -7\lambda + 6}$$

$$-2y = -6\lambda - 2\mu + 4, \qquad z = 2\mu + \lambda - 1$$

$$\boxed{-2y + z = -5\lambda + 3}$$

$$x - 3y = -\frac{7\,(-2y + z - 3)}{-5} + 6 + 5x - 15y$$

$$= 7(-2y + z - 3) + 30$$

$$5x - 15y + 14y - 7z + 21 - 30 = 0$$

$$\boxed{5x - y - 7z = 9}$$

(c) the cartesian form of the plane is therefore $\boxed{\mathbf{r}.(5\mathbf{i} - \mathbf{j} - 7\mathbf{k}) = 9}$

---

7. $\mathrm{d} = \dfrac{|\mathbf{b} \times (\mathbf{c} - \mathbf{a})|}{|\mathbf{b}|}$

$\mathbf{b} = -3\mathbf{i} + 4\mathbf{j} - \mathbf{k}$ the direction vector

$\mathbf{a} = 2\mathbf{i} + 3\mathbf{j} + 4\mathbf{k}$ the position vector

$\mathbf{c} = \mathbf{i} + 2\mathbf{j} + 3\mathbf{k}$ the direction vector of the point C

$|\mathbf{b}| = \sqrt{3^2 + 4^2 + 1^2} = \sqrt{26}$

$$\mathbf{c} - \mathbf{a} = \mathbf{i} + 2\mathbf{j} + 3\mathbf{k} - 2\mathbf{i} - 3\mathbf{j} - 4\mathbf{k} = -\mathbf{i} - \mathbf{j} - \mathbf{k}$$

$$\mathbf{b} \times (\mathbf{c} - \mathbf{a}) = (-3\mathbf{i} + 4\mathbf{j} - \mathbf{k}) \times (-\mathbf{i} - \mathbf{j} - \mathbf{k})$$

$$= \begin{vmatrix} \mathbf{i} & \mathbf{j} & \mathbf{k} \\ -3 & 4 & -1 \\ -1 & -1 & -1 \end{vmatrix} = \mathbf{i}(-4-1) - \mathbf{j}(3-1) + \mathbf{k}(3+4)$$

$$= -5\mathbf{i} - 2\mathbf{j} + 7\mathbf{k}$$

$$\mathrm{d} = \frac{|-5\mathbf{i} - 2\mathbf{j} + 7\mathbf{k}|}{\sqrt{26}} = \frac{\sqrt{25+4+49}}{\sqrt{26}} = \sqrt{3}.$$

---

8. (i) $$\int \underset{②}{\sinh^{-1} x}\ \underset{①}{\mathrm{d}x} = x \sinh^{-1} x - \int x \times \frac{1}{\left(1+x^2\right)^{\frac{1}{2}}}\,\mathrm{d}x$$

$$= x \sinh^{-1} x - \frac{1}{2}\int \frac{\mathrm{d}\left(1+x^2\right)}{\left(1-x^2\right)^{\frac{1}{2}}} = x \sinh^{-1} x - \left(1+x^2\right)^{\frac{1}{2}} + c$$

(ii) $$\int \underset{②}{\cosh^{-1} x}\ \underset{①}{\mathrm{d}x} = x \cosh^{-1} x - \int x \times \frac{1}{\sqrt{x^2-1}}\,\mathrm{d}x$$

$$= x \cosh^{-1} x - \frac{1}{2}\int \frac{\mathrm{d}\left(x^2-1\right)}{\left(x^2-1\right)^{\frac{1}{2}}} = x \cosh^{-1} x - \left(x^2-1\right)^{\frac{1}{2}} + c$$

(iii) $$\int \underset{②}{\tanh^{-1} x}\ \underset{①}{\mathrm{d}x} = x \tanh^{-1} x - \int x \times \frac{1}{1-x^2}\,\mathrm{d}x = x \tanh^{-1} x + \frac{1}{2}\int \frac{\mathrm{d}\left(1-x^2\right)}{1-x^2}$$

$$= x \tanh^{-1} x + \frac{1}{2}\ln\left|\left(1-x^2\right)\right| + c.$$

Examination Test Papers.

GCE Advanced Level.

Further Pure Mathematics FP3

Test Paper 8

Time: 1 hour and 30 minutes.

Instructions and Information

Candidates may use any calculator allowed by the regulations of their Examination Board.

Full marks are awarded for correct answers to ALL questions.

This paper has eight questions.

You can start working with any question and you must label clearly all parts.

1. Show that $\displaystyle\int_0^1 \frac{\sinh^2 x}{\cosh x}\,\mathrm{d}x = 1.175 - 2\tan^{-1} e + \frac{\pi}{2}$. **(10)**

---

2. Differentiate with respect to x :-

(i) $y = 3\sinh^{-1}\frac{1}{x}$ (ii) $y = \operatorname{cosech}^2\frac{x}{2}\operatorname{sech}^2\frac{x}{3}$. **(10)**

---

3. Show $\displaystyle\int x^p\left(\log_e x\right)^q \mathrm{d}x = \frac{x^{p+1}}{p+1}\left(\log_e x\right)^q - \frac{q}{p+1}\int x^p\left(\log_e x\right)^{q-1}\mathrm{d}x$

hence evaluate $\displaystyle\int_1^2 x^3\left(\log_e x\right)^3 \mathrm{d}x$. **(12)**

---

4. Determine the indefinite integral

$$\int \frac{1}{\sqrt{x^2 - 2x + 17}}\,\mathrm{d}x.$$ **(8)**

---

5. Determine the matrix **M** such that the point $P\ (x, y, z)$ is rotated through -90° (clockwise) about oz to the point $P'\left(x', y', z'\right)$ by the transformation

$$\mathbf{M}\begin{pmatrix} x \\ y \\ z \end{pmatrix} = \begin{pmatrix} x' \\ y' \\ z' \end{pmatrix}.$$ **(6)**

---

6. Find the distance of the point with position vector $\mathbf{i} + \mathbf{j} + \mathbf{k}$ from a plane π with equation $\mathbf{r}.\ (2\mathbf{i} + 3\,\mathbf{j} + 4\mathbf{k}) = 5$. **(9)**

---

7. Find the shortest distance between the two skew lines.

l_1: $\mathbf{r} = \mathbf{i} + \mathbf{j} + \lambda(2\mathbf{i} - \mathbf{j} + \mathbf{k})$

l_2: $\mathbf{r} = 2\mathbf{i} + \mathbf{j} - \mathbf{k} + \mu\,(3\mathbf{i} - 5\mathbf{j} + 2\mathbf{k})$. **(8)**

---

8. Find the locus of the curve whose parametric equations are $x = 4t^2$, $y = 4t$.

Hence find the length of the curve from $A(1, 2)$ to $B(4, 4)$. **(12)**

TOTAL FOR PAPER: 75 MARKS

Examination Test Papers.

GCE Advanced Level.

Test Paper 8 Solutions

Further Pure Mathematics FP3

1. $\int_0^1 \frac{\sinh^2 x}{\cosh x}\,dx = \int_0^1 \frac{\cosh^2 x - 1}{\cosh x}\,dx = \int_0^1 (\cosh x - \operatorname{sech} x)\,dx$

$= \int_0^1 \cosh x\,dx - \int_0^1 \operatorname{sech} x\,dx = \left[\sinh x\right]_0^1 - \int_0^1 \frac{2}{e^x + e^{-x}}\,dx$

$= 1.175 - \left[2\tan^{-1} e - \frac{\pi}{2}\right] = 1.175 - 2\tan^{-1} e + \frac{\pi}{2}$

where $\int_0^1 \frac{2}{e^x + e^{-x}}\,dx = \int_1^e \frac{2}{\left(u + \frac{1}{u}\right)}\frac{du}{u} = \int_1^e \frac{2}{u^2+1}\,du = \left[2\tan^{-1} u\right]_1^e$

let $u = e^x \Rightarrow \frac{du}{dx} = e^x \Rightarrow dx = \frac{du}{e^x} = \frac{du}{u}$

$\therefore \left[2\tan^{-1} u\right]_1^e = 2\tan^{-1} e - 2\tan^{-1} 1 = 2\tan^{-1} e - \frac{\pi}{2}$

$\therefore \int_0^1 \frac{\sinh^2 x}{\cosh x}\,dx = 1.175 - 2\tan^{-1} e + \frac{\pi}{2}.$

---

2. (i) $y = 3\sinh^{-1}\frac{1}{x} \Rightarrow \frac{1}{x} = \sinh\frac{y}{3} \quad x = \operatorname{cosech}\frac{y}{3} \quad \frac{dx}{dy} = \frac{1}{3}\left(-\operatorname{cosech}\frac{y}{3}\coth\frac{y}{3}\right)$

$\frac{dx}{dy} = -\frac{1}{3}\operatorname{cosech}\frac{y}{3}\coth\frac{y}{3} \quad \frac{dy}{dx} = -3\sinh\frac{y}{3}\tanh\frac{y}{3} = -\frac{3\sinh^2\frac{y}{3}}{\cosh\frac{y}{3}}$

$\frac{dy}{dx} = -\frac{3\left(\frac{1}{x}\right)^2}{\sqrt{1 + \left(\frac{1}{x}\right)^2}} = \frac{-\frac{3}{x^2}}{\sqrt{1 + \frac{1}{x^2}}} = -\frac{3}{x\sqrt{1+x^2}}$

(ii) $y = \operatorname{cosech}^2\frac{x}{2}\operatorname{sech}^2\frac{x}{3}$

$\frac{dy}{dx} = 2 \times \frac{1}{2}\operatorname{cosech}\frac{x}{2}\left(-\coth\frac{x}{2}\operatorname{cosech}\frac{x}{2}\right)\operatorname{sech}^2\frac{x}{3}$

$+ \operatorname{cosech}^2\frac{x}{2} \times \frac{2}{3}\operatorname{sech}\frac{x}{3}\left(-\tanh\frac{x}{3}\operatorname{sech}\frac{x}{3}\right)$

$= -\coth\frac{x}{2}\operatorname{cosech}^2\frac{x}{2}\operatorname{sech}^2\frac{x}{3} - \frac{2}{3}\operatorname{cosech}^2\frac{x}{2}\tanh\frac{x}{3}\operatorname{sech}^2\frac{x}{3}$

$= -\operatorname{cosech}^2\frac{x}{2}\operatorname{sech}^2\frac{x}{3}\left(\coth\frac{x}{2} + \frac{2}{3}\tanh\frac{x}{2}\right).$

---

3. $\int x^p \left(\log_e x\right)^q \, dx = \frac{x^{p+1}}{p+1}\left(\log_e x\right)^q - \frac{q}{p+1}\int x^p \left(\log_e x\right)^{q-1} \, dx$

$$\int \underset{①}{x^p} \underset{②}{\left(\log_e x\right)^q} \, dx = \frac{x^{p+1}}{p+1}\left(\log_e x\right)^q - \int \frac{x^{p+1}}{p+1} q \left(\log_e x\right)^{q-1} \frac{1}{x} \, dx$$

$$= \frac{x^{p+1}}{p+1}\left(\log_e x\right)^q - \int \frac{x^p}{p+1} q \left(\log_e x\right)^{q-1} \, dx$$

$$= \frac{x^{p+1}}{p+1}\left(\log_e x\right)^q - \frac{q}{p+1}\int x^p \left(\log_e x\right)^{q-1} \, dx$$

$p = 3, q = 3 \qquad \int_1^2 x^3 \left(\log_e x\right)^3 dx = \left[\frac{x^4}{4}\left(\log_e x\right)^3\right]_1^2 - \frac{3}{4}\int_1^2 x^3 \left(\log_e x\right)^2 dx$

$$\int_1^2 x^3 \left(\log_e x\right)^2 dx = \left[\frac{x^4}{4}\left(\log_e x\right)^2\right]_1^2 - \frac{2}{4}\int_1^2 x^3 \left(\log_e x\right) dx$$

$$\int_1^2 x^3 \left(\log_e x\right) dx = \left[\frac{x^4}{4}\left(\log_e x\right)\right]_1^2 - \frac{1}{4}\int_1^2 x^3 \, dx$$

$$\int_1^2 x^3 \, dx = \left[\frac{x^4}{4}\right]_1^2 = 4 - \frac{1}{4} = \frac{15}{4}$$

$$\therefore \int_1^2 x^3 \left(\log_e x\right)^3 dx = 4\left(\log_e 2\right)^3 - \frac{3}{4}\left[4\left(\log_e 2\right)^2 - \frac{1}{2}\left(4\log_e 2 - \frac{1}{4} \times \frac{15}{4}\right)\right]$$

$$= 4\left(\log_e 2\right)^3 - 3\left(\log_e 2\right)^2 + \frac{3}{8} \times 4\log_e 2 - \frac{3}{8} \times \frac{1}{4} \times \frac{15}{4}$$

$$= 1.33 - 1.44 + 1.04 - 0.352 = 0.578.$$

---

4. $\int \frac{1}{\left(x^2 - 2x + 17\right)^{\frac{1}{2}}} \, dx = \int \frac{dx}{\left[(x-1)^2 + 16\right]^{\frac{1}{2}}} = \int \frac{dx}{\left[(x-1)^2 + 4^2\right]^{\frac{1}{2}}}$

let $x - 1 = 4\sinh y \qquad \frac{dx}{dy} = 4\cosh y$

$$\int \frac{dx}{\left[(x-1)^2 + 4^2\right]^{\frac{1}{2}}} = \int \frac{1}{\left(16\sinh^2 y + 16\right)^{\frac{1}{2}}} 4\cosh y \, dy$$

$$= \int \frac{4\cosh y}{4\sqrt{\cosh^2 y}} \, dy = \sinh^{-1}\frac{x-1}{4} + c.$$

---

5. $\begin{pmatrix} 1 \\ 0 \\ 0 \end{pmatrix} \to \begin{pmatrix} 0 \\ -1 \\ 0 \end{pmatrix}, \quad \begin{pmatrix} 0 \\ 1 \\ 0 \end{pmatrix} \to \begin{pmatrix} 1 \\ 0 \\ 0 \end{pmatrix}, \quad \begin{pmatrix} 0 \\ 0 \\ 1 \end{pmatrix} \to \begin{pmatrix} 0 \\ 0 \\ 1 \end{pmatrix}.$

The required matrix is given

$$\mathbf{M} = \begin{pmatrix} 0 & 1 & 0 \\ -1 & 0 & 0 \\ 0 & 0 & 1 \end{pmatrix}.$$

---

6. $P = \mathbf{a} \,.\, \hat{\mathbf{n}} - D$

$$= (\mathbf{i} + \mathbf{j} + \mathbf{k}) \,.\, \frac{2\mathbf{i} + 3\mathbf{j} - 4\mathbf{k}}{\sqrt{2^2 + 3^2 + 4^2}} - \frac{5}{\sqrt{2^2 + 3^2 + 4^2}}$$

$$= (\mathbf{i} + \mathbf{j} + \mathbf{k}) \,.\, \frac{(2\mathbf{i} + 3\mathbf{j} + 4\mathbf{k})}{\sqrt{29}} - \frac{5}{29}$$

$$= \frac{2 + 3 + 4}{\sqrt{29}} - \frac{5}{\sqrt{29}} = \frac{4}{\sqrt{29}}$$

therefore the point P and the origin are on opposite sides of the plane.

---

7. $d = \left| (\mathbf{d} - \mathbf{c}) \,.\, \dfrac{\mathbf{b}_1 \times \mathbf{b}_2}{|\mathbf{b}_1 \times \mathbf{b}_2|} \right|$ the shortest distance between two skew lines.

$\mathbf{d} = \mathbf{i} + \mathbf{j} \qquad \mathbf{c} = 2\mathbf{i} + \mathbf{j} + \mathbf{k} \qquad \mathbf{b}_2 = 2\mathbf{i} - \mathbf{j} - \mathbf{k}$

$\mathbf{b}_1 = 3\mathbf{i} - 5\mathbf{j} + 2\mathbf{k}$

$$\mathbf{b}_1 \times \mathbf{b}_2 = \begin{vmatrix} \mathbf{i} & \mathbf{j} & \mathbf{k} \\ +3 & -5 & 2 \\ 2 & -1 & 1 \end{vmatrix} = \mathbf{i}(-5 + 2) - \mathbf{j}(3 - 4) + \mathbf{k}(-3 + 10) = -3\mathbf{i} + \mathbf{j} + 7\mathbf{k}$$

$$|\mathbf{b}_1 \times \mathbf{b}_2| = |-3\mathbf{i} + \mathbf{j} + 7\mathbf{k}| = \sqrt{9 + 1 + 49} = \sqrt{59} = 7.68$$

$$d = \left| \frac{(-\mathbf{i} + \mathbf{k}) \,.\, (-3\mathbf{i} + \mathbf{j} + 7\mathbf{k})}{7.68} \right|$$

$$= \frac{3 + 7}{7.58} = \frac{10}{7.58} = 1.30 \text{ units.}$$

---

8. $x = 4t^2,\ y = 4t \Rightarrow t = \frac{y}{4}, \qquad x = 4\left(\frac{y}{4}\right)^2 \qquad y^2 = 4x$ the locus.

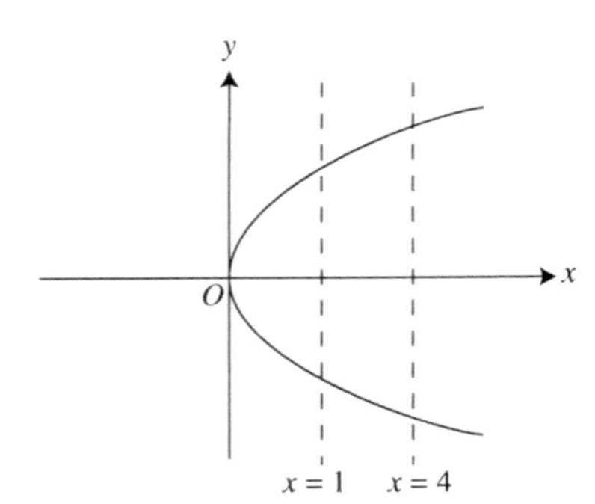

Differentiating with respect to x $\quad 2y\frac{dy}{dx} = 4 \quad \frac{dy}{dx} = \frac{4}{2y} = \frac{2}{y}.$

$$\text{Length of arc} = \int_1^4 \sqrt{1 + \left(\frac{dy}{dx}\right)^2}\,dx$$

$$= \int_1^4 \sqrt{1 + \frac{4}{y^2}}\,dx = \int_1^4 \sqrt{1 + \frac{4}{4x}}\,dx = \int_1^4 \sqrt{1 + \frac{1}{x}}\,dx$$

$$= \int_{\frac{1}{2}}^1 \sqrt{1 + \frac{1}{4t^2}}\,8t\,dt = \int_{\frac{1}{2}}^1 \frac{\sqrt{4t^2 + 1}}{2t}\,8t\,dt$$

$$= 4\int_{\frac{1}{2}}^1 \sqrt{4t^2 + 1}\,dt$$

where $x = 4t^2, \quad \frac{dx}{dt} = 8t, \quad x = 1, \quad t = \frac{1}{2}; \quad x = 4, \quad t = 1$

let $2t = \sinh y, \quad 2\frac{dt}{dy} = \cosh y$

$$4\int_{\frac{1}{2}}^1 \sqrt{4t^2 + 1}\,dt = \int_{\sinh^{-1} 1}^{\sinh^{-1} 2} 4\sqrt{\sinh^2 y + 1} \quad \cosh y\,\frac{dy}{2}$$

$$= 2\int_{\sinh^{-1} 1}^{\sinh^{-1} 2} \cosh^2 y\,dy = \int_{\sinh^{-1} 1}^{\sinh^{-1} 2} (\cosh 2y + 1)\,dy$$

$$= \left[\frac{\sinh 2y}{2} + y\right]_{\sinh^{-1} 1}^{\sinh^{-1} 2} = \frac{\sinh 2 \sinh^{-1} 2}{2} + \sinh^{-1} 2$$

$$-\frac{\sinh 2 \sinh^{-1} 1}{2} - \sinh^{-1} 1 = 3.62 \text{ units.}$$

Examination Test Papers.

GCE Advanced Level.

Further Pure Mathematics FP3

Test Paper 9

Time: 1 hour and 30 minutes.

Instructions and Information

Candidates may use any calculator allowed by the regulations of their Examination Board.

Full marks are awarded for correct answers to ALL questions.

This paper has eight questions.

You can start working with any question and you must label clearly all parts.

1. Evaluate the definite integral $\int_0^1 \sqrt{1+9x^2}\,dx$. **(8)**

---

2. Fig.1 shows the rectangular hyperbola with the parametric equations, $x = ct$, $y = \frac{c}{t}$.

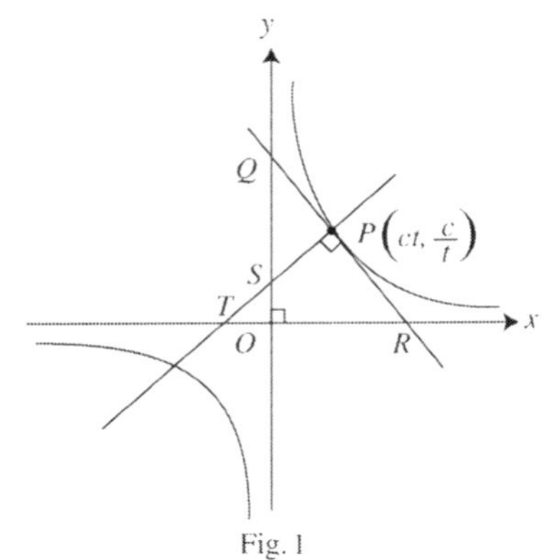

Fig. 1

Determine:

a) The equation of the curve.

b) The tangent and normal at $P\left(ct, \frac{c}{t}\right)$.

c) The coordinates at Q, R, S and T.

d) The areas of the triangles OQR and OST. **(12)**

---

3. Sketch the hyperbolas $xy = 1$ and $(x+1)(y-2) = 1$

and determine the coordinates of the points of intersections. **(10)**

---

4. Express $\cosh x + 2\sinh x$ in the form $R\sinh(x \pm \alpha)$ or $R\cosh(x \pm \alpha)$ giving α in logarithmic form. **(10)**

---

5. $\mathbf{M} = \begin{pmatrix} 1 & 0 & 0 \\ x & 2 & 0 \\ 3 & 1 & 1 \end{pmatrix}$.

Find $\mathbf{M}^{-1}$ in terms of x. **(8)**

---

6. (a) Find the perpendicular distance of the point $(1, -1, -1)$ from the plane

$$\mathbf{r}\,.\,(\mathbf{i}+\mathbf{j}+\mathbf{k}) = 1.$$

(b) Find the perpendicular distance of the point $(-1, 2, -3)$ from the plane

$$\mathbf{r}\,.\,(\mathbf{i}+\mathbf{j}-\mathbf{k}) = 2.$$

(8)

---

7. A line passes through a fixed point $A(4, 5, -7)$ and is parallel to a vector $2\mathbf{i} - 3\mathbf{j} + 5\mathbf{k}$. Determine:

(i) The vector equations of the line.

(ii) The parametric equations of the line.

(iii) The Cartesian equations of the line. **(9)**

---

8. Calculate the length of the arc of the curve whose parametric equations are $x = 3\cos t - \cos 3t$, $y = 3\sin t - \sin 3t$ between the points corresponding to $t = 0$ and $t = \frac{\pi}{4}$. Find first $\dot{x}$ and $\dot{y}$. **(10)**

TOTAL FOR PAPER: 75 MARKS

Examination Test Papers.

GCE Advanced Level.

Test Paper 9 Solutions

Further Pure Mathematics FP3

1. $\int_0^1 \sqrt{1+9x^2}\,dx = \int_0^{\sinh^{-1}3} \sqrt{1+\sinh^2 y}\,\frac{1}{3}\,dy \cosh y$

let $\sinh y = 3x$ $\quad \cosh y \frac{dy}{dx} = 3 \Rightarrow \frac{dy}{dx} = \frac{3}{\cosh y}$

$$\frac{1}{3}\int_0^{\sinh^{-1}3} \cosh^2 y \,\frac{dy}{dx} = \int_0^{\sinh^{-1}3} \frac{\cosh 2y + 1}{3 \times 2}\,dy = \frac{1}{6}\left[\frac{\sinh 2y}{2} + y\right]_0^{\sinh^{-1}3}$$

where $\cosh 2y = 2\cosh^2 y - 1$

$$= \frac{1}{6}\left[\frac{\sinh 2\sinh^{-1}3}{2} + \sinh^{-1}3\right]$$

$$= \frac{1}{6}(9.48683298 + 1.8184465)$$

$$= 1.88.$$

---

2. (a) $x = ct$ and $y = \frac{c}{t}$, eliminating t from these two equations

$$t = \frac{x}{c} \Rightarrow y = \frac{c}{\frac{x}{c}} = \frac{c^2}{x} \Rightarrow \boxed{xy = c^2}$$

(b) Differentiating (1) with respect to x

$$y + x\frac{dy}{dx} = 0 \Rightarrow \frac{dy}{dx} = -\frac{y}{x} = \frac{-\frac{c}{t}}{ct} = -\frac{1}{t^2}.$$

The equation of the tangent

$$y = -\frac{1}{t^2}\,x + k \Rightarrow \frac{c}{t} = -\frac{1}{t^2}\,ct + k \Rightarrow k = \frac{2c}{t}$$

$$y = -\frac{1}{t^2}\,x + \frac{2c}{t} \Rightarrow \boxed{t^2 y + x - 2ct = 0} \qquad \ldots(2)$$

The equation of the normal

$$y = t^2\,x + k \Rightarrow \frac{c}{t} = t^2\,ct + k \Rightarrow k = \frac{c}{t} - ct^3$$

$$y = t^2\,x + \frac{c}{t} - ct^3$$

$$\boxed{ty - t^3\,x + ct^4 = 0} \qquad \ldots(3)$$

(c) When $x = 0$, $\quad y = \dfrac{2c}{t}$ from (2)

$Q\left(0, \dfrac{2c}{t}\right)$ when $y = 0$, $\quad x = 2ct$ from (2) $\quad R(2ct, 0)$

when $x = 0$, $\quad y = ct^3$ from (3) $\quad$ when $y = 0$ $\quad x = ct$ from (3)

$S\left(0, ct^3\right) \quad T\,(ct, 0)$

(d) $\triangle OQR \quad \dfrac{2c}{t} \times 2ct \times \dfrac{1}{2} = 2c^2 \quad \triangle OST \quad \dfrac{ct^3 \times ct}{2} = \dfrac{c^2 t^4}{2}.$

---

3. $xy = 1 \quad \ldots (1)$ and $(x+1)(y-2) = 1 \quad \ldots (2)$

Solving the equations (1) and (2) we have

$(x+1)(y-2) = 1$

$xy + y - 2x - 2 = 1 \Rightarrow 1 + y - 2x - 2 = 1 \Rightarrow y - 2x = 2$

$y - 2x = 2 \Rightarrow y = 2x + 2 \Rightarrow xy = x(2x+2) = 1 \Rightarrow \boxed{2x^2 + 2x - 1 = 0}$

$x = \dfrac{-2 \pm \sqrt{4+8}}{2 \times 2} = -\dfrac{1}{2} \pm \dfrac{\sqrt{3}}{2} \quad x = -\dfrac{1}{2} + \dfrac{\sqrt{3}}{2}$ or $x = -\dfrac{1}{2} - \dfrac{\sqrt{3}}{2}$

and the corresponding values of y are

$y = 2\left(-\dfrac{1}{2} + \dfrac{\sqrt{3}}{2}\right) + 2 = 1 + \sqrt{3}$ or $y = 2\left(-\dfrac{1}{2} - \dfrac{\sqrt{3}}{2}\right) + 2 = -\sqrt{3} + 1$

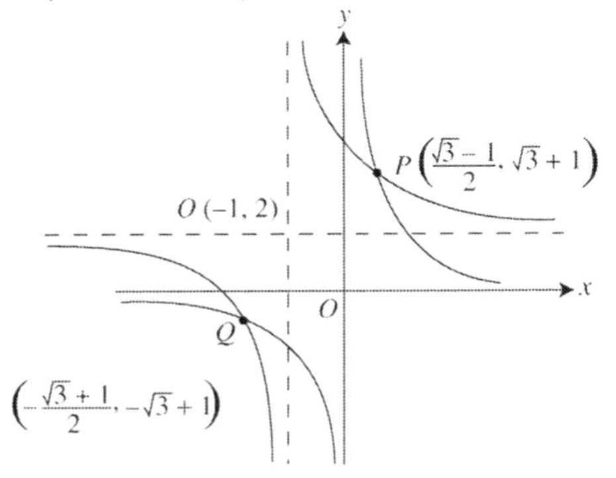

---

4. $2\sinh x + \cosh x \equiv R\sinh(x+\alpha)$

$$\equiv R\sinh x\cosh\alpha + R\sinh\alpha\cosh x$$

$R\sinh\alpha = 1 \qquad \ldots(1)$

$R\cosh\alpha = 2 \qquad \ldots(2)$

Squaring up both sides

$R^2\sinh^2\alpha = 1 \qquad \ldots(3)$

$R^2\cosh^2\alpha = 4 \qquad \ldots(4)$

$(4)-(3)$

$R^2\left(\cosh^2\alpha - \sinh^2\alpha\right) = 4-1=3 \qquad R=\sqrt{3}$

Dividing (1) by (2)

$\tanh\alpha = \frac{1}{2}$ but $\tanh^{-1}x = \frac{1}{2}\ln\frac{1+x}{1-x}$

$$\alpha = \tanh^{-1}\frac{1}{2} = \frac{1}{2}\ln\frac{1+\frac{1}{2}}{1-\frac{1}{2}} = \frac{1}{2}\ln 3 = \ln\sqrt{3}$$

$\therefore 2\sinh x + \cosh x = \sqrt{3}\sinh\left(x+\ln\sqrt{3}\right).$

---

5. $|\mathbf{M}| = \begin{vmatrix} 1 & 0 & 0 \\ x & 2 & 0 \\ 3 & 1 & 1 \end{vmatrix} = 1\begin{vmatrix} 2 & 0 \\ 1 & 1 \end{vmatrix} = 2$

$$\mathbf{M}^* = \begin{pmatrix} 2 & -x & x-6 \\ 0 & 1 & -1 \\ 0 & 0 & 2 \end{pmatrix} \quad \mathbf{M}^{*\mathrm{T}} = \begin{pmatrix} 2 & 0 & 0 \\ -x & 1 & 0 \\ x-6 & -1 & 2 \end{pmatrix}$$

$$\mathbf{M}^{-1} = \frac{\mathbf{M}^{*\mathrm{T}}}{|\mathbf{M}|} = \frac{1}{2}\begin{pmatrix} 2 & 0 & 0 \\ -x & 1 & 0 \\ x-6 & -1 & 2 \end{pmatrix}$$

$$= \begin{pmatrix} 1 & 0 & 0 \\ -\frac{x}{2} & \frac{1}{2} & 0 \\ \frac{x-6}{2} & -\frac{1}{2} & 1 \end{pmatrix}.$$

---

6. (a) $p = \mathbf{a} \,.\, \widehat{\mathbf{n}} - D$

$$= (\mathbf{i} + \mathbf{j} - \mathbf{k}) \,.\, \frac{(\mathbf{i} + \mathbf{j} + \mathbf{k})}{\sqrt{3}} - \frac{1}{\sqrt{3}}$$

$$= \frac{-2}{\sqrt{3}} \quad \text{the point and the origin are on the same side.}$$

(b) $p = \mathbf{a} \,.\, \widehat{\mathbf{n}} - D = (-\mathbf{i} + 2\mathbf{j} - 3\mathbf{k}) \,.\, \dfrac{\mathbf{i} + \mathbf{j} - \mathbf{k}}{\sqrt{3}} - \dfrac{2}{\sqrt{3}}$

$$= -\frac{1}{\sqrt{3}} + \frac{2}{\sqrt{3}} + \frac{3}{\sqrt{3}} - \frac{2}{\sqrt{3}} = \frac{2}{\sqrt{3}}$$

the point and the origin are on opposite sides of the plane.

---

7. (i) $\mathbf{r} = \mathbf{a} + \lambda \mathbf{b}$ the vector equation of the line, $\mathbf{r} = (4\mathbf{i} + 5\mathbf{j} - 7\mathbf{k}) + \lambda(2\mathbf{i} - 3\mathbf{j} + 5\mathbf{k})$ or

$\mathbf{r} = (4 + 2\lambda)\mathbf{i} + (5 - 3\lambda)\mathbf{j} + (-7 + 5\lambda)\mathbf{k}$.

(ii) The parametric equations of the line

$\mathbf{r} = x\mathbf{i} + y\mathbf{i} + z\mathbf{k} = (4 + 2\lambda)\mathbf{i} + (5 - 3\lambda)\mathbf{j} + (-7 + 5\lambda)\mathbf{k}$

$$x = 4 + 2\lambda \qquad \ldots (1)$$

$$y = 5 - 3\lambda \qquad \ldots (2)$$

$$z = -7 + 5\lambda \qquad \ldots (3)$$

The parametric equations of the line.

(iii) The Cartesian equations of the line are found by equating λ in each case of (1), (2) and (3)

$$\lambda = \frac{x - 4}{2} = \frac{y - 5}{-3} = \frac{z + 7}{5}.$$

---

8. Length of arc = $\displaystyle\int_0^{\frac{\pi}{4}} \sqrt{\dot{x}^2 + \dot{y}^2}\, \mathrm{d}t$

$x = 3\cos t - \cos 3t \qquad \dfrac{\mathrm{d}x}{\mathrm{d}t} = -3\sin t + 3\sin 3t = \dot{x}$

$y = 3\sin t - \sin 3t \qquad \dfrac{\mathrm{d}y}{\mathrm{d}t} = 3\cos t - 3\cos 3t = \dot{y}$

Length of arc

$$= \int_0^{\frac{\pi}{4}} \left(9\sin^2 t + 9\sin^2 3t - 18\sin t \sin 3t + 9\cos^2 t + 9\cos^2 3t - 18\cos t \cos 3t\right)^{\frac{1}{2}} \mathrm{d}t$$

$$= \int_0^{\frac{\pi}{4}} \sqrt{9 + 9 - 18\left(\cos t \cos 3t + \sin t \sin 3t\right)}\, \mathrm{d}t$$

$$= \sqrt{18} \int_0^{\frac{\pi}{4}} \sqrt{1 - 1 + 2\sin^2 t}\, \mathrm{d}t \qquad \cos 2t = 1 - 2\sin^2 t$$

$$= \sqrt{18}\sqrt{2} \int_0^{\frac{\pi}{4}} \sin t \,\mathrm{d}t = 6 \int_0^{\frac{\pi}{4}} \sin t \,\mathrm{d}t = \left[-6 \cos t\right]_0^{\frac{\pi}{4}}$$

$$= \frac{-6\sqrt{2}}{2} + 6 = \left(6 - 3\sqrt{2}\right) \text{ units.}$$

Examination Test Papers.

GCE Advanced Level.

Further Pure Mathematics FP3

Test Paper 10

Time: 1 hour and 30 minutes.

Instructions and Information

Candidates may use any calculator allowed by the regulations of their Examination Board.

Full marks are awarded for correct answers to ALL questions.

This paper has eight questions.

You can start working with any question and you must label clearly all parts.

1. Evaluate $\int_{0.5}^{1} \operatorname{cosech} x \, dx$. **(6)**

2.

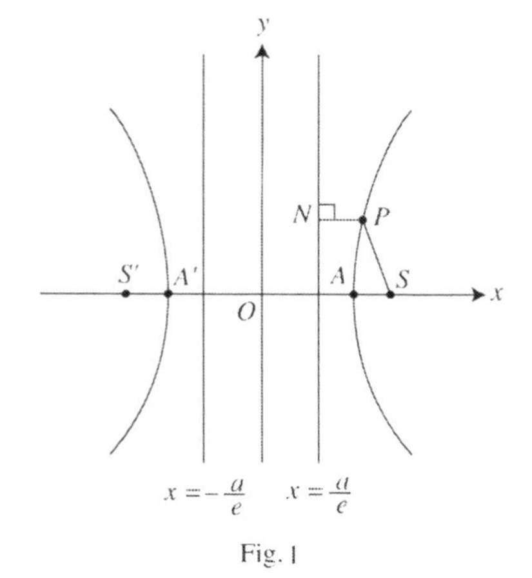

Fig. 1

The equation of the asymptotes are shown and the equation of the curve is $\frac{x^2}{a^2} - \frac{y^2}{b^2} = 1$.

a) Write down the coordinates of the foci and those intersecting the x - axis, and the relationship between a and b.

b) Define the locus of this curve in terms of the eccentricity, stating the value of e.

c) State the trigonometric and hyperbolic parametric equations.

d) Show that $y = \pm \frac{b}{a} x$ are the other asymptotes and sketch them. **(12)**

---

3. Find the equations of the tangent and normal at the point $P\ (a \sec \theta, b \tan \theta)$ of the curve $\frac{x^2}{a^2} - \frac{y^2}{b^2} = 1$. **(11)**

---

4. Evaluate $\int_0^4 \frac{dx}{12 \sinh x + 5 \cosh x}$. **(10)**

---

5. The position vectors of the points P, Q, and R relative to an origin O is respectively

$\overrightarrow{OP} = 3\mathbf{i} + 6\mathbf{k}, \quad \overrightarrow{OQ} = 5\mathbf{j} + 3\mathbf{k}, \quad \overrightarrow{OR} = \mathbf{i} + \mathbf{k}.$

Find (a) the area of ΔPQR,

(b) an equation of the plane PQR in the form $\mathbf{r} \cdot \mathbf{n} = k$, where $k \in \mathbb{R}$.

The transformation represented by the matrix $\mathbf{M}$ where $\mathbf{M} = \begin{pmatrix} 2 & 1 & 0 \\ 1 & -1 & 1 \\ 5 & 1 & 0 \end{pmatrix}$ maps the points A, B, C to the points P, Q, R respectively. **(12)**

---

6. Determine whether the points $(1, -2, 1)$, $(-2, 1, 3)$ are on the same or opposite sides of the plane $\quad \mathbf{r} \cdot (\mathbf{i} + 2\mathbf{j} + \mathbf{k}) = 1$. **(6)**

---

7. Write down the general equation of a plane.

Find the equation of the plane for the point $A(0, -3, -7)$ on the plane which is perpendicular to the line with direction ratios $1:2:3$. **(6)**

---

8. An asteroid is given by the Fig. 2

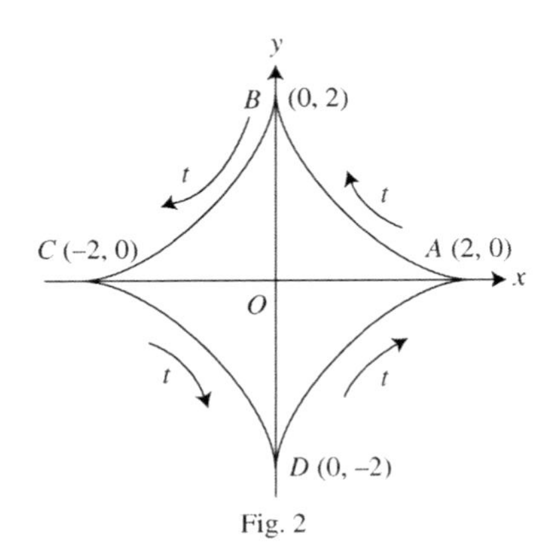

Fig. 2

and the parametric equations

$$x = 2\cos^3 t, \qquad y = 2\sin^3 t \qquad 0 \le t \le 2\pi.$$

a) Determine the length of arc $0 \le t \le \frac{\pi}{2}$, hence find the perimeter $ABCD$ of the asteroid.

b) The arc AB is totated about the x- axis, find the surface are so formed and hence find the surface area if the asteroid. **(12)**

TOTAL FOR PAPER: 75 MARKS

Examination Test Papers.

GCE Advanced Level.

Test Paper 10 Solutions

Further Pure Mathematics FP3

1. $\displaystyle\int_{0.5}^{1} \operatorname{cosech} x \, \mathrm{d}x = \int_{0.5}^{1} \frac{1}{\sinh x} \, \mathrm{d}x = 2\int_{0.5}^{1} \frac{1}{e^x - e^{-x}} \, \mathrm{d}x$

let $u = e^x \quad \dfrac{\mathrm{d}y}{\mathrm{d}x} = e^x, \, \mathrm{d}x = \dfrac{\mathrm{d}y}{e^x}$

$$2\int_{0.5}^{1} \frac{1}{e^x - e^{-x}} \, \mathrm{d}x = 2\int_{e^{0.5}}^{e} \frac{1}{u - \frac{1}{u}} \, \frac{\mathrm{d}u}{e^x} = 2\int_{e^{.0.5}}^{e} \frac{\mathrm{d}u}{u^2 - 1}$$

$$= \int_{0.5}^{e} \left(\frac{1}{u-1} - \frac{1}{u+1} \right) \mathrm{d}u \quad \text{where } \frac{1}{u^2-1} = \frac{1}{(u-1)(u+1)} = \frac{\frac{1}{2}}{u-1} - \frac{\frac{1}{2}}{u+1}$$

$$\int_{e^{0.5}}^{e} \left(\frac{1}{u-1} - \frac{1}{u+1} \right) \mathrm{d}u = \left[\ln(u-1) - \ln(u+1) \right]_{e^{0.5}}^{e}$$

$$= \ln \frac{e-1}{e+1} - \ln \frac{e^{0.5}-1}{e^{0.5}+1} = \ln 0.462 - \ln 0.245 = 0.634$$

2.

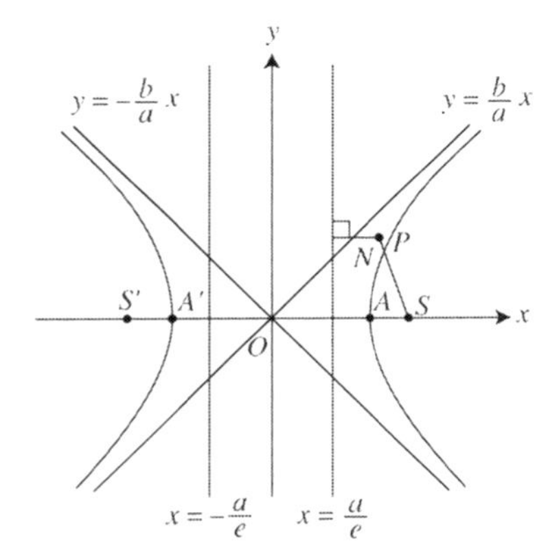

a) $A(a, 0) \qquad S(ae, 0) \qquad A'(-a, 0) \qquad S'(-ae, 0) \qquad b^2 = a^2\left(e^2 - 1\right)$

b) $e = \dfrac{SP}{PN} > 1$

c) $x = a\sec\theta, \; y = b\tan\theta$ trigonometric

$x = a\cosh\theta, \; y = b\sinh\theta$ hyperbolic

d) $\dfrac{x^2}{a^2} - \dfrac{y^2}{b^2} = 1 \Rightarrow b^2x^2 - a^2y^2 = a^2b^2 \qquad \ldots(1)$

if $y = mx + c$ is tangent to (1)

$$b^2x^2 - a^2(mx + c)^2 = a^2b^2$$

$$b^2x^2 - a^2m^2x^2 - a^2c^2 - 2mxca^2 - a^2b^2 = 0$$

$$x^2\left(b^2 - a^2m^2\right) - 2mca^2x - \left(a^2c^2 + a^2b^2\right) = 0 \qquad \ldots (2)$$

the straight line, $y = mx + c$, is tangent when the discriminant of (2) is zero

$$D = b^2 - 4ac = \left(-2mca^2\right)^2 + 4\left(a^2c^2 + a^2b^2\right)\left(b^2 - a^2m^2\right) = 0$$

$$4m^2c^2a^4 + 4a^2c^2b^2 + 4a^2b^4 - 4a^4c^2m^2 - 4a^4b^2m^2 = 0$$

$$4a^2b^2\left(c^2 + b^2 - a^2m^2\right) = 0$$

$$c = a^2m^2 - b^2$$

Solving (2) by putting $D = 0$ we have

$$x = \frac{2a^2cm}{2\left(b^2 - a^2m^2\right)} = \frac{a^2cm}{b^2 - a^2m^2} \text{ but } c = \pm\sqrt{a^2m^2 - b^2}$$

$$x = \frac{a^2m}{-c} \Rightarrow x = \pm\frac{a^2m}{\sqrt{a^2m^2 - b^2}} \qquad \text{when } m \to \pm\frac{b}{a}, \quad c = 0, x \to \pm\infty$$

$$\therefore \boxed{y = \pm\frac{b}{a}x}$$

3.

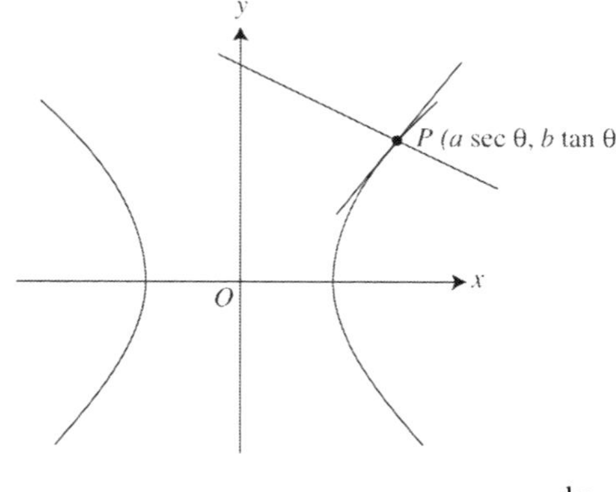

$$x = a\sec\theta$$

$$\frac{\mathrm{d}y}{\mathrm{d}x} = a\sec\theta\tan\theta$$

$$y = b\tan\theta, \quad \frac{\mathrm{d}y}{\mathrm{d}\theta} = b\sec^2\theta$$

$$\frac{\frac{\mathrm{d}y}{\mathrm{d}\theta}}{\frac{\mathrm{d}x}{\mathrm{d}\theta}} = \frac{\mathrm{d}y}{\mathrm{d}x} = \frac{b\sec^2\theta}{a\sec\theta\tan\theta} = \frac{b}{a}\operatorname{cosec}\theta$$

The gradient of the tangent at $P\ (a\sec\theta, b\tan\theta)$ is $\frac{b}{a}\operatorname{cosec}\theta$

$y = \left(\frac{b}{a}\operatorname{cosec}\theta\right)x + c$, which passes through the point P,

$$b\tan\theta = \left(\frac{b}{a}\operatorname{cosec}\theta\right)a\sec\theta + c \qquad c = b\tan\theta - b\operatorname{cosec}\theta\sec\theta$$

$$\therefore y = \left(\frac{b}{a}\operatorname{cosec}\theta\right)x + b\left(\tan\theta - \operatorname{cosec}\theta\sec\theta\right)$$

multiplying each term by $a\sin\theta$

$$ay\sin\theta = bx + ab\left(\frac{\sin^2\theta}{\cos\theta} - \sec\theta\right)$$

$$ay\sin\theta = bx + ab\,\frac{(\sin^2\theta - 1)}{\cos\theta} = bx + ab\,(-\cos\theta)$$

$$\boxed{bx - ay\sin\theta - ab\cos\theta = 0}$$

The gradient of the normal is $-\frac{a}{b}\sin\theta$ and the equation of the normal

$y = -\frac{a}{b}(\sin\theta)\,x + c$, this passes through P.

$$b\tan\theta = -\frac{a}{b}(\sin\theta)\,a\,(\sec\theta) + c \Rightarrow c = b\tan\theta + \frac{a^2}{b}\frac{\sin\theta}{\cos\theta}$$

$$y = -\frac{a}{b}(\sin\theta)\,x + b\tan\theta + \frac{a^2}{b}\sin\theta\sec\theta$$

$$\therefore yb + ax\sin\theta - b^2\tan\theta - a^2\sin\theta\sec\theta = 0$$

$$\boxed{ax\sin\theta + yb - \left(a^2 + b^2\right)\tan\theta = 0}$$

---

4. $\int_0^1 \frac{dx}{12\sinh x + 5\cosh x}$

$$12\sinh x + 5\cosh x \equiv R\sinh x\cosh\alpha + R\sinh\alpha\cosh x = R\sinh(x + \alpha)$$

$R\cosh\alpha = 12 \qquad \ldots(1)$

$R\sinh\alpha = 5 \qquad \ldots(2)$

$$\tan\alpha = \frac{5}{12} \Rightarrow \tanh^{-1}\frac{5}{12} = \frac{1}{2}\ln\frac{1+\frac{5}{12}}{1-\frac{5}{12}} \qquad \alpha = \frac{1}{2}\ln\frac{17}{7}$$

$$R^2\cosh^2\alpha - R^2\sinh^2\alpha = 12^2 - 5^2 = 144 - 25 = 119 \qquad R = \sqrt{119}$$

$$12\sinh x + 5\cosh x = \sqrt{119}\,\sinh\left(x + \frac{1}{2}\ln\frac{17}{7}\right)$$

$$\int_0^1 \frac{\mathrm{d}x}{12\sinh x + 5\cosh x} = \int_0^1 \frac{\mathrm{d}x}{\sqrt{119}\sinh\left(x + \ln\sqrt{\frac{17}{7}}\right)}$$

$$= \frac{1}{\sqrt{119}}\int_0^1 \operatorname{cosech}\left(x + \ln\frac{\sqrt{17}}{\sqrt{7}}\right)\mathrm{d}x = \frac{1}{\sqrt{119}}\left[\ln\tanh\left(\frac{x}{2} + \ln\frac{\sqrt{17}}{\sqrt{7}}\right)\right]_0^1$$

$$= \frac{1}{\sqrt{119}}\left[\ln\tanh\left(\frac{x}{2} + 1.56\right)\right]_0^1 = \frac{1}{\sqrt{119}}\left[\ln\tanh(0.5 + 1.56) - \ln\tanh 1.56\right]$$

$$= \frac{1}{\sqrt{119}}(-0.032491887 + 0.088371804) = 0.005122503 = 0.005 \text{ to 3 d.p.}$$

---

5. $\overrightarrow{PQ} = -3\mathbf{i} + 5\mathbf{j} - 3\mathbf{k}, \quad \overrightarrow{QR} = \mathbf{i} - 5\mathbf{j} - 2\mathbf{k} \quad \overrightarrow{RP} = 2\mathbf{i} + 5\mathbf{k}$

(a) Area $\Delta PQR = \frac{1}{2}\left|\overrightarrow{PQ} \times \overrightarrow{QR}\right| = \frac{1}{2}\begin{vmatrix} \mathbf{i} & \mathbf{j} & \mathbf{k} \\ -3 & 5 & -3 \\ 1 & -5 & -2 \end{vmatrix}$

$$= \frac{1}{2}\left|-25\mathbf{i} - 9\mathbf{j} + 10\mathbf{k}\right| = \frac{1}{2}\sqrt{25^2 + 9^2 + 10^2}$$

$$= \frac{1}{2}\sqrt{806} = 14.2 \text{ to 3 s.f.}$$

(b) A normal to the plane PQR is $\mathbf{n} = \overrightarrow{PQ} \times \overrightarrow{QR}$ $\mathbf{n} = \overrightarrow{PQ} \times \overrightarrow{QR} = -25\mathbf{i} - 9\mathbf{j} + 10\mathbf{k}$.

Therefore, the equation of the plane is $\mathbf{r}.(-25\mathbf{i} - 9\mathbf{j} + 10\mathbf{k}) = k$ where $k \in \mathbb{R}$

where the position vector $\overrightarrow{OP}$ is $3\mathbf{i} + 6\mathbf{k}$

$\therefore k = (3\mathbf{i} + 6\mathbf{k}).(-25\mathbf{i} - 9\mathbf{j} + 10\mathbf{k}) = -75 + 60 = -15.$

Therefore the plane ABC has equation

$\mathbf{r}.(-25\mathbf{i} - 9\mathbf{j} + 10\mathbf{k}) = -15$ or $\mathbf{r}.(25\mathbf{i} + 9\mathbf{j} - 10\mathbf{k}) = 15$

(c) $|\mathbf{M}| = \begin{vmatrix} 2 & 1 & 0 \\ 1 & -1 & 1 \\ 5 & 1 & 0 \end{vmatrix} = (-1)(2-5) = 3$ by expanding down column 3.

$$\mathbf{M}^{*\mathrm{T}} = \begin{pmatrix} -1 & 0 & 1 \\ 5 & 0 & -2 \\ 6 & 3 & -3 \end{pmatrix} \qquad \mathbf{M}^{-1} = \frac{\mathbf{M}^{*\mathrm{T}}}{|\mathbf{M}|} = \frac{1}{3}\begin{pmatrix} -1 & 0 & 1 \\ 5 & 0 & -2 \\ 6 & 3 & -3 \end{pmatrix}.$$

Let $\mathbf{X}$ be the matrix whose column are the position of A, B and C and $\mathbf{Y}$ be the matrix whose columns are the position vectors of the corresponding mapped points P, Q, R.

$\therefore \mathbf{MX} = \mathbf{Y} \Rightarrow \mathbf{X} = \mathbf{M}^{-1}\mathbf{Y} \Rightarrow \mathbf{X} = (A\ B\ C)$

$$\mathbf{X} = \frac{1}{3}\begin{pmatrix} -1 & 0 & 1 \\ 5 & 0 & -2 \\ 6 & 3 & -3 \end{pmatrix}\underset{P\quad Q\quad R}{\begin{pmatrix} 3 & 0 & 1 \\ 0 & 5 & 1 \\ 6 & 3 & 0 \end{pmatrix}} \Rightarrow \mathbf{X} = \frac{1}{3}\begin{pmatrix} 3 & 3 & -1 \\ 3 & -6 & 5 \\ 0 & 6 & 9 \end{pmatrix}$$

$$\underset{A\quad B\quad C}{\begin{pmatrix} 1 & 1 & -\frac{1}{3} \\ 1 & -2 & \frac{5}{3} \\ 0 & 2 & 3 \end{pmatrix}}$$

$\therefore$ A has position vector $\mathbf{i} + \mathbf{j}$

B has position vector $\mathbf{i} - 2\mathbf{j} + 2\mathbf{k}$

C has position vector $-\frac{1}{3}\mathbf{i} + \frac{5}{3}\mathbf{j} + 3\mathbf{k}$.

---

6. $p = \mathbf{a}\,.\,\widehat{\mathbf{n}} - D = (\mathbf{i} - 2\mathbf{j} + \mathbf{k})\,.\,\dfrac{(\mathbf{i} + 2\mathbf{j} - \mathbf{k})}{\sqrt{6}} - \dfrac{1}{\sqrt{6}}$

$$= \frac{1}{\sqrt{6}} - \frac{4}{\sqrt{6}} - \frac{1}{\sqrt{6}} - \frac{1}{\sqrt{6}} = -\frac{5}{\sqrt{6}}$$

$$p = \mathbf{a}\,.\,\widehat{\mathbf{n}} - D = (-2\mathbf{i} + \mathbf{j} + 3\mathbf{k})\,.\,\frac{\mathbf{i} + 2\mathbf{j} - \mathbf{k}}{\sqrt{6}} - \frac{1}{\sqrt{6}}$$

$$= -\frac{2}{\sqrt{6}} + \frac{2}{\sqrt{6}} - \frac{3}{\sqrt{6}} - \frac{1}{\sqrt{6}} = -\frac{4}{\sqrt{6}}$$

the points are on the same side of the plane since they are both of the same sign and the points and the origin are on the same side.

7.

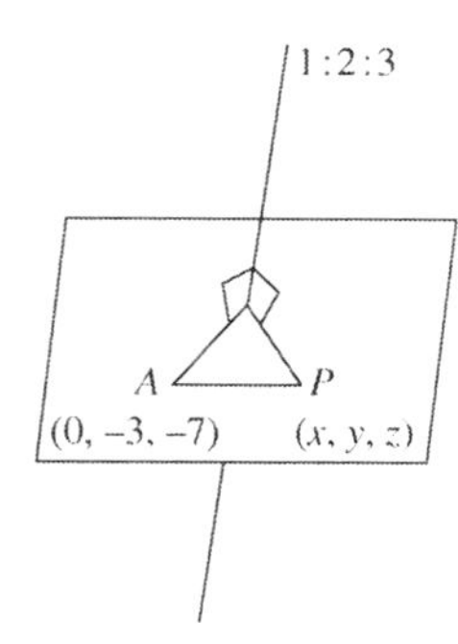

$ax + by + cz = d$

The direction ratios of AP are $x : (y+3) : (z+7)$

and since the line is perpendicular to

AP $1(x) + 2(y+3) + 3(z+7) = 0$

$x + 2y + 6 + 3z + 21 = 0$

$\boxed{x + 2y + 3z = -27}$

the equation of the plane and $P(x, y, z)$ is a point on it.

---

8. (a) $$\int_0^{\frac{\pi}{2}} \sqrt{\left(\frac{dx}{dt}\right)^2 + \left(\frac{dy}{dt}\right)^2}\, dt = \int_0^{\frac{\pi}{2}} \sqrt{(36 \sin^2 t \cos^4 t + 36 \sin^4 t \times \cos^2 t)}dt$$

$$x = 2\cos^3 t, \quad \frac{dx}{dt} = 6\cos^2 t(-\sin t) \qquad y = 2\sin^3 t, \quad \frac{dy}{dt} = 6\sin^2 t \cos t$$

$$= 6\int_0^{\frac{\pi}{2}} \sin t \cos t \sqrt{\cos^2 t + \sin^2 t}\, dt = \int_0^{\frac{\pi}{2}} 3 \sin 2t\, dt$$

$$= \left[-\frac{3}{2}\cos 2t\right]_0^{\frac{\pi}{2}} = -\frac{3}{2}\cos \pi + \frac{3}{2}\cos 0° = 3 \text{ units.}$$

The perimeter of the astroid $= 4 \times 3 = 2$ units

(b) $$\int_0^{\frac{\pi}{2}} 2\pi y\,(3\sin 2t)\, dt = \int_0^{\frac{\pi}{2}} 2\pi\, 2\sin^3 t\; 3\sin 2t\, dt = 2 \times 12\pi \int_0^{\frac{\pi}{2}} \sin^4 t\, \cos t\, dt$$

$$= 24\pi \int_0^{\frac{\pi}{2}} \sin^4 t\, d(\sin t) = 24\pi \left(\frac{\sin^5 t}{5}\right)_0^{\frac{\pi}{2}} = \frac{24\pi}{5}$$

Total surface area $= \dfrac{48\pi}{5}$.

Lightning Source UK Ltd.
Milton Keynes UK
UKOW031215141212

203584UK00001B/17/P